The
New Mexico
Natural History Series

Barry S. Kues, Editor

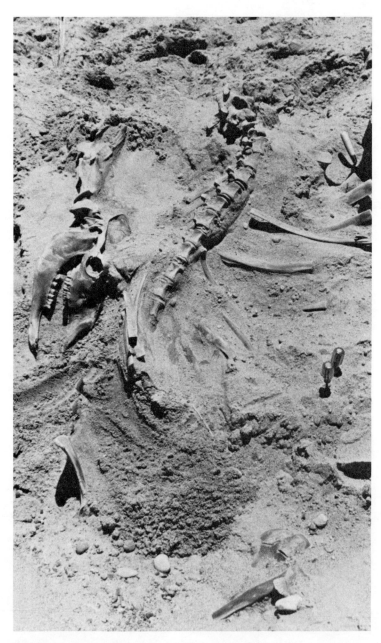

Excavation of Pleistocene camel skeleton from a gravel quarry in Albuquerque.

FOSSILS of NEW MEXICO

Barry S. Kues

New Mexico Natural History Series

University of New Mexico Press
Albuquerque

Library of Congress Cataloging in Publication Data

Kues, Barry S.
  Fossils of New Mexico.

  (The New Mexico natural history series)
  Bibliography: p.
  Includes index.
  1. Paleontology—New Mexico.   I. Title.   II. Series.
QE747.N6K83   560'.9789   82-4864
ISBN 0-8263-0622-5   AACR2
ISBN 0-8263-0613-6 (pbk.)

© 1982 by the University of New Mexico Press.
All rights reserved.
Manufactured in the United States of America.
International Standard Book Number (clothbound) 0-8263-0622-5
International Standard Book Number (paperbound) 0-8263-0613-6
Library of Congress Catalog Card Number 82-4864
Second paperbound printing 1986

This book is dedicated to my wife, Georgianna, who endured patiently and supportively my long days and evenings at the office during its writing.

CONTENTS

	Preface	xiii		Precambrian	86
1	**Basic Principles**	3		Cambrian	87
	The Nature of Fossils and Paleontology	4		Ordovician	89
	Fossilization	5		Silurian	96
	Types of Fossilization	7		Devonian	98
	Where Fossils Are Found	11		Mississippian	102
	The Geologic Time Scale	16		Pennsylvanian	105
	Uses of Fossils	20		Permian	121
	Collecting Fossils	23		Triassic	134
	Specimen Preparation	30		Jurassic	140
				Cretaceous	148
2	**The Main Fossil Groups**	33		Paleocene	176
	Classification	33		Eocene	183
	The Main Divisions of Life	35		Oligocene	190
	Animals and Animal Fossils—Some General Features	37		Miocene-Pliocene	191
				Quaternary	196
	Invertebrate fossils	41		The Present	206
	Vertebrate fossils	62	4	**Where to Go for Further Information**	211
	Plants	73		In General	211
3	**The New Mexico Fossil Record**	83		References	212
	Distribution of Fossiliferous Rocks in New Mexico	84		**Index**	217

vii

ILLUSTRATIONS

Excavation of Pleistocene camel skeleton, Albuquerque	ii
Petrified wood	8
A carbonized Pennsylvanian fern	9
A mold and a steinkern	10
Fossil burrows	10
Shale	11
Sandstone	12
Limestone	13
Footprints of a Pleistocene camel, near Santa Fe	15
Part of a topographic map	24
Part of a geologic map	25
Collecting equipment	27
Modes of life in the ocean	38
Types of symmetry	39
Solitary and colonial organization	40
Protozoa	42
Porifera	43
Coelenterata	44
Bryozoa	46
Brachiopoda	48
Annelida	49
Gastropoda	51
Pelecypoda	52
Cephalopoda and minor mollusc classes	54
Arthropoda	57
Echinodermata	59
Graptolites	60
Conodonts	61
Relationships of major groups of aquatic vertebrates	64
Relationships of major groups of reptiles and reptile derivatives	67
Relationships of major groups of archosaurs	70
Relationships of major mammal groups	72
A Late Silurian scene	76
A Pennsylvanian coal forest	77
Lower vascular plants	79
Gymnosperms and angiosperms	80
Piece of a stromatolite	86
Cambrian fossils	88
Reconstruction of a Late Ordovician marine environment	91
Ordovician fossils	92
Ordovician fossils	93
Ordovician fossils	94
Silurian fossils	97
Devonian fossils	99
Devonian fossils	100
Mississippian shallow marine environment	104
Mississippian fossils	106
Mississippian fossils	107
Mississippian fossils	108
Fusulinids, near Jemez Springs	109
Branch of *Walchia*, Manzano Mountains	111

ILLUSTRATIONS

Part of trunk of *Lepidodendron*	112
Front portion of *Adelophthalmus*, near Los Lunas	113
Pennsylvanian shallow marine environment	114
Pennsylvanian fossils	115
Pennsylvanian fossils	116
Pennsylvanian fossils	117
Pennsylvanian fossils	118
Pennsylvanian fossils	119
Chaetetes, near Jemez Springs	120
Cross section through rocks of Permian reef complex	123
Reconstruction of marine life, Permian reef complex	124
Reconstruction of *Dimetrodon*	126
Reconstruction of *Limnoscelis*	127
Reconstruction of *Ophiacodon*	128
Skeleton of *Eryops*	128
Skeleton of *Sphenacodon*	128
Reptile footprints, Fra Cristobal Mountains	129
Permian fossils	130
Permian fossils	131
Permian fossils	132
Permian fossils	133
Skulls of *Metoposaurus*	136
Skeletal reconstruction of *Metoposaurus*	137
Reconstruction of a Late Triassic environment	138
Triassic fossils	139
Cast of *Coelophysis* skeleton	140
Skeletal reconstruction of *Coelophysis*	141
A Jurassic fish	143
Model of *Stegosaurus*	145
Reconstruction of *Antrodemus*	145
Skeleton of *Camarasaurus*	146
Paleontologist excavating *Camarasaurus* vertebra	146
Camarasaurus vertebra after laboratory preparation	146
Cretaceous fossils	151
Cretaceous fossils	152
Cretaceous fossils	153
Cretaceous fossils	154
Cretaceous fossils	155
Cretaceous fossils	156
Cretaceous fossils	157
Reconstruction of a Late Cretaceous seafloor	158
Coilopoceras	160
Acanthoceras	160
Ophiomorpha	160
Mosasaur skeleton	161
A palm frond impression	162
A petrified log in the Bisti badlands	163
"Fossil forest," northwestern New Mexico	164
Turtle shell	165
Garfish scales	166
Crocodilian scutes	166
A crocodilian tooth	166
Skeletal restoration of *Parasaurolophus*	167
Skeleton and restoration of *Pentaceratops*	168
Skull and lower jaw of *Kritosaurus*	169
Ceratopsian limb bone	170
Ankylosaur plate	171
Carnosaur tooth	172
Tyrannosaurus rex skeleton	173
Part of vertebral column, *Kritosaurus*	174
Kritosaurus skeleton, during excavation	175
Part of *Allognathosuchus* jaw	178
Akanthosuchus limb	179
Small mammal jaw	180
Claenodon skull	181
Restoration of *Ptilodus*	183
Turtle shell fragment	186
Coryphodon jaw	187
Restoration of *Coryphodon*	188
Restoration of *Phenacodus* and *Meniscotherium*	188
Hyracotherium jaw fragment	189
Restoration of *Hyracotherium*	189
Restoration of *Diatryma*	190
Bottom of skull of *Teleodus*	191
Partial jaw of a rhinocerous	194

ILLUSTRATIONS

Partial jaw of a merycodont	194	Restoration of the "Columbian mammoth" and *Smilodon*	202
Lower jaw fragment of Miocene horse	194	Skeleton of *Camelops*	203
Skull of doglike carnivore	195	Restoration of *Nothrotheriops* and *Smilodon*	204
Land tortoise	195	Paleo-Indian spear point	205
Early Pliocene landscape	197	*Coilopoceras*	208
Part of mammoth jaw	200		
Mastodon molars	201		

MAPS

Outcrops of Cambrian, Ordovician, and Silurian rocks
   in New Mexico   90
Outcrops of Devonian and Mississippian rocks in New Mexico   103
Outcrops of Pennsylvanian rocks in New Mexico   110
Outcrops of Permian rocks in New Mexico   122
Outcrops of Triassic rocks in New Mexico   135
Outcrops of Jurassic rocks in New Mexico   142
Outcrops of Cretaceous rocks in New Mexico   149
Outcrops of Paleocene rocks in New Mexico   177
Outcrops of Eocene rocks in New Mexico   184
Outcrops of Miocene-Pliocene rocks in New Mexico   192
Outcrops of Quaternary age in New Mexico   198

TABLES

1   The geologic time scale   18
2   Distribution and extent of rocks and fossils in New Mexico   85

PREFACE

Fossils collected from New Mexico sedimentary rocks have long been studied by geologists and paleontologists, and information about their nature and occurrence is contained in more than 2,000 scientific papers. Fossils, however, have also been collected by, and are of interest to, those with no formal training in paleontology. This book has been written primarily for them. It provides a nontechnical introduction to New Mexico's abundant and diverse fossils, so that one who has found some fossils may gain an understanding of the kinds of organisms they represent, their age, the environment in which they lived, and their significance to our understanding of the long and complex history of life. Discussion and illustrations of the more important and common types of New Mexico fossils found in rocks of different ages in the state are included in order to provide the reader a means of identifying the fossils he or she has collected, without having to search out and use the technical literature. An understanding of New Mexico fossils will increase one's appreciation and enjoyment of them, and it is my hope that this book will serve that purpose.

A book summarizing the fossils of New Mexico could not have been written without the previous efforts of the scores of paleontologists and geologists who have studied New Mexico's fossil record for the past 140 years. Much of the information and many of the illustrations in this book have been taken from their work. Especially valuable in providing illustrations of New Mexico fossils utilized in making the line drawings were the *Treatise on Invertebrate Paleontology* and Shimer and Shrock's *Index Fossils of North America,* along with numerous technical papers on various aspects of the state's paleontology. I also thank the Smithsonian Institution, which provided several of the photo-

PREFACE

graphs of ancient environments and permission to use them in the book, Jay Matternes, for allowing reproduction of his painting of Pliocene mammals, and Professor Z. Spinar and Artia of Prague, Czechoslovakia, who kindly gave me permission to reproduce several of Z. Burian's magnificent paintings of ancient animals and plants. Professor Edwin Colbert provided assistance in obtaining several illustrations of Triassic fossils.

I am grateful also to Spencer Lucas, Yale University, and Stuart Northrop, University of New Mexico, for reading the manuscript and offering many helpful suggestions on how it could be improved.

FOSSILS of NEW MEXICO

Do you wonder that the paleontologist walks a little apart from the ways of men . . . ? His field of vision embraces the whole of life. His time scale is so gigantic that it dwarfs to insignificance the centuries of human endeavor. And the laws and principles which he studies are those which control the whole great stream of life, upon which the happenings of our daily existence appear but as surface ripples.

W. D. Matthew

1

BASIC PRINCIPLES

Fossils are the remains of ancient plants and animals naturally preserved in the rocks of the earth's crust. These remains are often parts of the structure of the organism (a shell or bone), but may also represent the behavior of the organism (a trackway or burrow), or even the fragmentary or altered chemical constituents of a once-living creature. They are almost invariably found in sedimentary rocks, for reasons discussed later, and are abundant in the rocks throughout New Mexico. Our state is fortunate in having a fossil record that includes each geologic period from the Cambrian (500+ million years ago) to the Recent (the last 10,000 years), and nearly every imaginable ancient environment. Numerous New Mexico fossil deposits are of national and international importance, and nearly 1,000 different kinds of fossils were originally made known to the scientific world from specimens first found in New Mexico rocks. In fact, some of New Mexico's fossils have never been found anywhere outside of the state. New Mexico's fossils are indeed a varied and highly significant natural resource—a resource that is constantly becoming better understood through continuous new discoveries by paleontologists and laymen alike. One of the most exciting aspects of collecting and learning about the state's fossils is that new types keep turning up, found not only by paleontologists and geologists, but by amateur collectors ranging from children to college students to senior citizens. Many of these subjects will be explored at greater length farther on, but before considering New Mexico's fossils specifically it is desirable to have in mind some general principles concerning the nature of fossils, where they are found, and what they are used for.

BASIC PRINCIPLES

The Nature of Fossils and Paleontology

Fossils have been collected and examined since prehistoric times. Some tribes of American Indians collected fossils to use as beads and other ornaments, and medicinal powers were ascribed to certain fossils such as ammonoids. Fossils appear in the writings of the ancient Greeks; Herodotus, examining fossilized foraminifers in Egypt around 450 B.C., suggested that they were the petrified remains of uneaten lentils that had been dropped by workers building the pyramids! A few Greeks, noting the presence of shells high in the mountains of their country, hypothesized that the sea had once covered the land in those areas, but this was definitely a minority view, far ahead of its time.

The concept that a fossilized shell or plant leaf represents part of a once-living organism seems obvious to us. A fossil leaf, for example, shows the same kinds of features that we can easily observe on modern ones. Many petrified clam shells look almost identical to ones we can find on the beach. But strangely, it has been only within the last 200 years that the correct interpretation of the nature of fossils has been widely accepted. Before that, during medieval and renaissance times, a great variety of other explanations were advanced. Fossils were considered to be the result of gaseous precipitations deep within the earth, or the remains of heavenly bodies that had fallen from the sky, or tricks of the devil. The original meaning of the word *fossil* was anything dug up out of the earth; thus one of the first books to illustrate fossils (*On Fossil Objects*, by Konrad von Gesner, 1565) was mostly devoted to gemstones, mineral ores, and useful rocks such as marble and slate.

As true fossils began to be more extensively collected and studied, the undeniable similarities of many fossils to living organisms could no longer be seriously challenged, and by about 1800, the idea that fossils were the petrified remains of ancient organisms was accepted by most scientists. As more and more fossils were observed, however, it became apparent that some of them represented plants and animals unknown in the modern world. The idea that organisms could become extinct was difficult for many people to accept, because it implied that there were imperfections in God's natural order. Rather than suppose that some fossils represented extinct organisms, it was popular to believe that apparently vanished creatures still lived in remote parts of the world that mankind had not yet explored. Thomas Jefferson, third president of the United States and an amateur naturalist, when confronted with a fossil mammoth tooth was certain that living mammoths could be found in the vast forests of the western United States, and specifically requested that Lewis and Clark look for them during their lengthy explorations in the Louisiana Purchase.

With the acceptance that fossils were the remains of often extinct animals and plants, and with the emerging concept of the earth being millions, rather than a few thousands, of years old, the true scientific study of fossils, or paleontology, began. Since that time, with the critical realization that different types of fossils lived at different times in the past, paleontology has played an important role in the founding of the science of geology, and today is a multifaceted discipline. Within the general discipline of paleontology are such specialized subdivisions as *invertebrate paleontology*, the study of ancient invertebrate animals; *vertebrate paleontology*, the study of ancient vertebrate animals; *paleobotany*, the study of ancient plants; *paleoecology*, the study of the interrelationships of ancient organisms with each other and with their environment; *paleobiogeography*, the study of the past geographic distribution of animals and plants; *micropaleontology*, the study of microscopic fossils; *biostratigraphy*, the study of the exact time relationships of fossils and the strata they are found in; *palynology*, the study of fossil spores and pollen; *taphonomy*, the study of how fossils are deposited.

Fossilization

Though it may not seem obvious to one who is standing in an area where thousands of fossils are visible within, or eroding out of, a rock outcrop, fossilization of an ancient organism is a rare and unusual event. Only a minute fraction of organisms that ever lived have become fossilized, and only a very small percentage of those will ever be available for human observation and study. To understand why this is so, let us consider what may happen to a typical animal on the way to becoming fossilized, and as an example, let us use a fish that lived in the great shallow sea that covered New Mexico about 100 million years ago. The first step in the fossilization process is the death of the animal. Death can come in many forms, but one of the most common is being eaten by some other animal. If the predator completely digests the prey, there will be little or nothing left of the fish to become fossilized. If the fish escapes being eaten, it dies of other causes and settles

BASIC PRINCIPLES

for a time on the surface of the seafloor. Here it may be subjected to ingestion by scavengers, but if it escapes this, the soft parts (muscles, skin, internal organs) will begin to decay, and the body will gradually become covered by sediments. Decay will result in the eradication of the soft parts, removing those parts of the fish's structure from preservation as a fossil, and leaving only the skeleton, which is composed of hard mineralized material. In shallow water, the possibility of the skeleton being broken into small unrecognizable fragments by currents, waves, or the activities of other organisms is significant, again removing the fish from the possibility of becoming fossilized.

A small fraction of the total dead fish in the New Mexico sea may escape these processes and become gently buried deep enough in the sediments so that other organisms or waves and currents can no longer affect them. As sediment builds up over the remaining skeletons, they become compressed, and eventually the sediment begins to lithify, or turn to rock. Chemical changes associated with lithification may destroy the potential fossil, or it may endure, deep within the developing sedimentary rock, the molecules of its structure being transformed or replaced by other molecules derived from the rock or introduced in fluids that percolate between the sedimentary grains. The skeleton is becoming fossilized.

Eventually the marine sediments are uplifted, and the sedimentary rock is raised above sea level. Fossilization, if not complete, will continue. If the sedimentary rock is subjected to volcanic activities, or to extreme heat and pressure at depth, however, the incipient fossils within may be destroyed. If the rock becomes part of the land, and is eroded, the fossils it contains will probably be destroyed. And even if the fossil escapes destruction, only a small proportion of the volume of the earth's fossil-bearing rocks has been exposed at the earth's surface during the time when mankind has been around to collect the fossils these rocks contain.

As a result of the processes outlined above, only a small percentage of ancient organisms will survive relatively intact to become fossilized; only a small percentage of fossils will escape subsequent destruction by geologic processes; and only a small percentage of those fossils that do survive will ever be available for human inspection. The next time you find a fossil, regard it with more than a little awe; its presence in the rocks is the chance result of its escaping many processes that could have destroyed it over the millions or hundreds of millions of years since it was a living organism.

Types of Fossilization

The actual fossilization process may occur in several different ways, and the appearance of a particular specimen will in part be controlled by its mode of fossilization. We can divide fossils into three general groups: (1) *Body Fossils,* in which all or part of the shape and structure of the organism is preserved; (2) *Trace Fossils,* which are patterns or structures in the sediment caused by the behavior of an animal; and (3) *Chemical Fossils,* in which no evidence of shape, structure, or behavior is present, but some of the chemical composition of the organism is preserved. These modes of fossilization are discussed in more detail below.

In body fossils, the original materials out of which the organism was constructed have almost always been altered during fossilization. Rarely, only the soft parts decay, with no change in the skeletal elements, but usually there is change in the mineralized skeleton as well. Even more rarely, the organism is preserved under special conditions in which no alteration of either the hard or soft parts occurs. The various types of body fossil preservation follow.

1. Unaltered. Occasionally animals die in situations where little or no decay of the soft parts occurs. Perhaps the best examples are the numerous Pleistocene mammoths found frozen in the tundra of Siberia and Alaska. Because of the extreme cold in these regions, these extinct beasts are preserved absolutely intact; the meat can be thawed out and used to feed sled dogs and the undigested stomach contents can be analyzed to determine what the last meal was. Preservation in very dry environments (mummification) may also result in little or no change in an animal, aside from the loss of its body fluids. An excellent example of mummified fossils is the extinct ground sloth found by youths exploring some volcanic crevices near Las Cruces in the 1920s; preservation was so good that the fur was intact and the stomach contents could be identified.

2. Unaltered Hard Parts. Organisms with relatively stable mineralized or organic skeletons may become fossilized without alteration of their hard parts; only the soft parts are lost to decay. Many invertebrates with shells of calcite endure without significant chemical change to these shells, and even soft-bodied animals, such as insects preserved in amber (fossilized tree sap), occasionally are preserved this way.

3. Recrystallization. Often during fossilization, the original shell may be gradually dissolved, the minerals going into solution, to

BASIC PRINCIPLES

Petrified wood. The original woody material has been replaced by SiO_2 (silica), and original cavities have been filled in to produce a brightly colored, hard fossil. In this example, petrifaction has not preserved the original cell structure of the wood, but often such fine structures are preserved.

be nearly simultaneously reprecipitated. The shell is "remade" in a sense, out of the same chemical materials it was composed of originally, but the crystal form of the reprecipitated mineral may be different, and the crystals are often larger than they were in the original shell.

4. *Replacement.* Frequently fossilization proceeds by the gradual dissolution of the original skeletal mineral and the replacement of the original substance by some other mineral. If this happens slowly, the replacement will occur on a molecule-by-molecule basis, producing a perfect replica of the original skeleton out of the new material. The most common type of replacement involves $CaCO_3$ (the mineral calcite or aragonite) shells being replaced by SiO_2 (the mineral quartz). Pennsylvanian fossils in the Jemez Springs area are sometimes a bright red color, due to the replacement of the original calcite shell by agate, a form of quartz. Uncommonly, other minerals may do the replacing; some fossils are composed of iron pyrite (fool's gold) and have a heavy golden metallic feel to them; a few fossils are replaced with iron oxide or more exotic minerals such as garnet, native silver, and fluorite.

5. *Petrifaction.* When the original skeletal material of a fossil has been replaced, and, in addition, the pores and cavities of the skeleton have been filled by the replacement mineral, the fossil is said to be petrified. The petrified wood so common in northwestern New Mexico is a good example of this type of fossilization; not only has the cellulose of the wood been replaced by quartz (often forms of quartz with bright coloration, like jasper and agate), but the cells themselves have been filled, creating a completely solid, dense fossil. As with replacement, petrifaction usually occurs over long periods of time, under the influence of ground water percolating through the sediments. Minerals dissolved in

TYPES OF FOSSILIZATION

A carbonized Pennsylvanian fern. The fossil is a thin carbon layer that preserves some of the details of the original plant.

the ground water are deposited within the skeleton and cavities of the organisms contained within the sediments. In some cases, large cavities may be present, like the decayed center of a log, and minerals are precipitated as crystals around the periphery of the cavity, creating a geode. Petrifaction is perhaps the main way in which fossil bones are formed; minerals replace the original calcium phosphate of the bone and also fill in all of the small cavities within the bone.

6. *Carbonization.* Under some conditions organisms lacking mineralized skeletons are preserved as black fossils, often flattened on bedding planes of the rock. The organic material of the dead organism has been altered by chemical processes, leading to the breakdown (but not complete destruction) of its complex organic molecules. The nitrogen, oxygen, hydrogen, and other atoms that are typical constituents of organic molecules leave, but the carbon in the molecules remains. Under ideal conditions the shape, structure, and fine detail of the organism are preserved, but sometimes the result is an unidentifiable carbonaceous smudge. Large volumes of plant remains, under the proper physical and chemical conditions, may be transformed into the carbonized fossils we know as coal.

7. *Molds, Casts, and Steinkerns.* If you press a seashell firmly into clay and then remove the shell, you have the impression of the outer surface of the shell left in the clay. Many fossils are formed in an analogous way; a shell or other part of an organism makes an impression on the sediment either before or after burial, and then the shell itself is dissolved, leaving a mold as the only indication of the shell's former presence. When a buried shell is dissolved chemically, a cavity is left that is outlined by the mold of the shell. If, later on, minerals derived from ground water pre-

BASIC PRINCIPLES

Left: A mold and a steinkern. A2 is an indentation in the rock that preserves the original ornamentation of a clam shell (mold); A1 is the sedimentary infilling of another clam that remained after the shell had dissolved away (steinkern). Right: Fossil burrows. Filled-in worm and crustacean burrows are among the most common types of trace fossils.

cipitate in the cavity and fill it, a cast of the shell is produced. Fossils with large internal cavities, such as snails, clams, and brachiopods, have these cavities filled with sediment as burial occurs. The shape of these cavity-fillings corresponds to the shape of the cavity, and if the shell is dissolved, the filling, or steinkern, may be the only remains of the organism left.

Trace Fossils are very common types of fossils, particularly those formed by the movement of a worm or other invertebrate through the sediments. These burrows may become filled with sediment and preserved—testaments to the activity of an ancient animal that reveal little about the nature or structure of the animal that made it. The footprints of a dinosaur or other vertebrate in sediments are also trace fossils, reflecting the behavior of the animal while it was living. Similarly, fossil feces (coprolites) and fossil eggs are also considered trace fossils.

Chemical Fossils give us the least amount of information about the nature of the organisms they represent, because they are merely organic molecules that once formed part of an animal or plant. Their presence in the sediments can be detected with the appropriate analytical devices, and information derived from chemical fossils can be useful. The presence of chlorophyll molecules in very old rocks reveals that plants existed then, because only plants manufacture these molecules; but little about the actual nature of the plants or what they looked like can be deduced. Petroleum is a chemical fossil—the organic molecules that compose petroleum must have derived from past organisms, but scientists are still uncertain as to exactly what ancient creatures were involved. Chemical fossils are mainly studied by geochemists rather than paleontologists, and will not be considered further in this book.

Shale, with carbonized remains of the pelecypod *Dunbarella*, from the Pennsylvanian of the Manzano Mountains. Shale is essentially lithified mud, with grains too small to see with the naked eye. Shale outcrops often erode into slopes or rounded hills, and the fossils in them may be easily collected as they become disassociated from the rock.

Where Fossils Are Found

Virtually all fossils are found in sedimentary rocks. Sedimentary rocks are formed at low temperatures and pressures at or near the earth's surface by the transportation, deposition, and eventual accretion of grains of material originally eroded from previously existing rocks, or as a result of precipitation of minerals from water. Among the commonest sedimentary rocks are familiar types such as shale (grains smaller than 1/16 mm, usually too fine to see with the naked eye), sandstone (grains 1/16 to 2 mm), conglomerate (grains 2 mm to pebble size or larger), and limestone (formed from $CaCO_3$ grains of all sizes either precipitated directly from seawater or precipitated by organisms as part of their shells). Because sedimentary rocks are formed by processes acting all over the earth's surface, they are quite widespread, amounting to 70 per-

BASIC PRINCIPLES

Sandstone. The sand grains of the original sediment have become cemented by matrix, and are large enough to be observed with the naked eye. Sandstone beds may be very hard, forming ledges, or relatively soft and easily erodable, depending on the matrix.

cent or more of the rocks exposed in an area such as the state of New Mexico. And because the remains of organisms are affected by the same processes that lead to the transportation and accumulation of sedimentary grains, it is not surprising that almost all sedimentary rocks contain fossils.

The degree to which a given sedimentary rock will contain fossils, and the types of fossils likely to be found in it, depend largely on the environment in which the sediments (and fossils) were originally deposited. As with modern organisms, ancient plants and animals existed in, and were deposited in a variety of different environments. Generally fossils are more abundant in rocks deposited in marine environments than they are in terrestrial rocks. Partly this is because in the past, as today, oceanic environments were present over the great majority (about 70 percent) of the earth's surface, and marine sedimentary rocks are thus

Limestone. This slab of Pennsylvanian limestone consists entirely of pieces of the calcareous skeletons of marine invertebrates, mostly brachiopods and crinoid stems. Most limestones are relatively hard and massive, and their fossils are difficult to extract, but some erode readily and may yield thousands of isolated fossils.

far more abundant than terrestrial sedimentary rocks. An equally important consideration is the fact that marine environments are predominantly depositional—sediments are deposited and accumulate in them—whereas terrestrial environments for the most part are erosional. Organisms dying in a marine environment stand a reasonably good chance of being buried and ultimately becoming fossilized, but an organism dying on land generally will not, unless it is washed into a stream, lake, or other body of water into which sediments are being deposited. These freshwater bodies, however, are really very small compared to marine environments. Finally, many marine organisms are relatively sessile, many actually being attached to the substrate, and possessing hard, compact, massive skeletons that can resist fragmentation and fossilize relatively easily. Most terrestrial organisms, on the other hand, have skeletons that become disarticulated and scattered after death (such as vertebrates), or have skeletons that either are not mineralized or are soft and poorly mineralized (plants, insects, worms, and so forth), and are thus much less amenable to fossilization.

The grain size of the enclosing sediment strongly affects the potential for fossilization and the quality of preservation. In coarse-grained sediments, such as a sandy beach deposit, movement of the grains by currents or waves causes abrasion of the remains of organisms within the sediment, sometimes gradually pulverizing shells before they become buried. As environments in which coarse-grained sediments are deposited are often characterized by high levels of water agitation, shells may become fragmented as well as abraded, leading to poor or no preservation. Deposits containing very large pebbles (conglomerates) usually characterize highly active environments (a fast-moving stream, for example) and only the largest and most resistant skeletal parts are likely to endure to become fossilized. Fine-grained muds, on the other hand, such as are deposited in quiet offshore marine, or in mudflat or swamp, environments, are ideal for the preservation of fossils, not only because the grains do not damage skeletal elements, but also because a fine-grained substrate may actually preserve detailed impressions of the soft parts of organisms. Most of our knowledge of ancient soft-bodied organisms, such as insects, worms, and plant leaves, comes from deposits of shales and mudstones.

Limestones almost invariably contain fossils of one kind or another; in fact, many limestones are composed entirely of fossil shell fragments. Most limestones form in warm, shallow marine environments, where there is a profusion of life, most of it possessing $CaCO_3$ skeletons. Today, in places like the Florida Keys or around the great Pacific and Caribbean reefs, enormous volumes of calcareous sediment are forming, as generations of shelled organisms continually contribute their skeletons to the sediment. As warm, marine, carbonate-depositing environments were far more widespread at some times in the past than they are today, the abundance of fossiliferous limestones is not surprising.

The above discussion has briefly indicated some of the factors that affect the distribution of fossils in the earth's sedimentary rocks. Added to these, of course, is the simple principle that sedimentary environments in which large numbers of organisms lived will yield more fossils than environments in which few organisms lived, all other factors being equal. Most fossils one is likely to find are of organisms that lived in ancient shallow marine environments, not only because these environments were extensive

Fossil Pleistocene camel tracks in a volcanic ash layer near Santa Fe. The tracks were probably made when the ash was wet, and they were subsequently covered by a thin layer of black cinders from a nearby volcanic vent. Left photo shows a view of part of the trackway; plaster casts have just been made of two of the footprints. Right photo shows closeup of footprints. Tracks such as these are uncommon in sedimentary rocks; their occurrence in a volcanic setting is highly unusual. These footprints were covered over a few years ago by the B.L.M. to prevent their destruction by erosion or vandalism.

in the past and represent excellent sedimentological conditions for preservation, but also because they have been optimum habitats for abundant and diverse populations of animals (and to a lesser degree plants) through at least the last 600 million years of earth history.

Few fossils are preserved in nonsedimentary rocks. Igneous rocks, formed from extremely hot melts originating deep within the earth, very rarely have fossils, because the high temperatures are sufficient to burn, vaporize, or otherwise completely destroy any organism coming into contact with the molten igneous material. Fossils found in igneous rocks include charred incompletely burned bones, the impressions of tree trunks made in a relatively cool, viscous lava that flowed through a forest, and, in cases where lava or ash had cooled enough to become almost solidified, animal footprints. Footprints of ancient camels are known in a volcanic ash deposit near Santa Fe, but such occurrences are very rare.

Metamorphic rocks, formed by the alteration of existing rocks under conditions of high temperature and pressure, also rarely have fossils. Only in cases where fossiliferous sedimentary rocks have been subjected to relatively mild temperatures and pressures do the contained fossils survive metamorphic alteration. Generally, fossils in metamorphic rocks are distorted and recrystallized, usually to the extent that little information may be gained from them. Geologists and paleontologists look only in sedimentary rocks for fossils, because the possibility of finding fossils in igneous or metamorphic rocks is extremely remote.

The Geologic Time Scale

Living organisms have been a part of the earth's history since shortly after the planet itself was formed; the earth is about 4.5 billion years old and the earliest known fossils are at least 3.5 billion years old. Fossils have played a critical role in our understanding of the history of the earth and in providing the means of determining very precisely the timing of events in the last 600 million years of this history.

The geologic time scale was developed in the nineteenth century as a direct result of an explosion in paleontological studies. As knowledge of the distribution of fossils in sedimentary rocks increased, geologists realized that specific fossils and fossil assemblages were confined to limited sequences of rock. Various fossils that were common and widespread in some strata gave way to different assemblages in overlying (younger) rocks. By constantly comparing fossils from each newly studied stratigraphic section with previously studied assemblages, a grand and detailed picture of the succession of different organisms through the sedimentary rock record emerged. This succession extended from the very oldest fossiliferous rocks known at the time, bearing strange and primitive-looking fossils, to rocks holding fossils of plants and animals nearly identical to ones we see in the world today.

It was further recognized that rocks bearing the same kinds of fossils represented the same interval of time, even if the rocks were separated by thousands of kilometers and were on different continents. Names were given to intervals of sedimentary rocks having generally similar and characteristic types of fossils, and boundaries between the intervals were defined at points where changes in the fossils were striking and abrupt. These names were based on localities where the fossils seemed to be particularly well displayed (for example, Devonian, for Devon, England; or Permian, for the city of Perm, in Russia), or on distinctive features of the rocks that held the fossils (for example, Cretaceous, meaning chalk).

This definition of a succession of intervals, each with characteristic fossils, formed the basis of the geologic time scale in use internationally today (see Table 1). There have been minor adjustments in the time scale, and there are some variations in the main intervals used from place to place. North Americans, for example, recognize two periods, Mississippian and Pennsylvanian, in the interval where geologists on other continents recognize only the single Carboniferous period. As more and more of the world's sedimentary rocks have been studied, it has also become apparent that the faunal and floral changes from period to period are not necessarily as abrupt as was once believed. And, as knowledge of fossils has increased, it has become possible to divide each period into many smaller subdivisions, in some cases as many as fifty or more.

By about 1900, the main features of the time scale were well established. The scale, however, was a strictly relative one. Silurian fossils, for instance, wherever found could be identified as such immediately, and it was known that they were younger than Ordovician fossils, but older than Devonian fossils. The absolute age of Silurian fossils—how many millions of years before the present they had lived—was, on the other hand, anybody's guess. The advent of radiometric dating in the early twentieth century, based on the rate and degree of decay of certain radioactive isotopes to different daughter elements, allowed determination of the absolute age of each part of the time scale.

Punctuating the time scale are three places where very profound changes in the world's plants and animals occurred: at the beginning of the Cambrian, when complex shelled multicellular animals appeared suddenly in the record; and at the end of the Permian and Cretaceous periods, when massive extinctions of previously dominant organisms occurred, with subsequent replacement by much different dominant forms. These places form the boundaries between the four largest divisions of earth history—the Precambrian, Paleozoic, Mesozoic, and Cenozoic.

The geologic time scale, then, forms the temporal framework within which we determine the order of succession and the age of fossils and rocks. Knowing the age of a fossil assemblage allows

BASIC PRINCIPLES

Table 1. The geologic time scale.
Numbers are millions of years before the present and indicate the beginning date for a geologic period or epoch.

Age (Ma)	Era	Period	Epoch	Description
	CENOZOIC	Quaternary	Recent	Evolution of civilized man, who becomes the most dominant biological factor.
.01				
			Pleistocene	Extensive glaciations; many large mammals flourished, becoming extinct at end of epoch; evolution of modern man.
2				
		Tertiary	Pliocene	Continued diversification of essentially modern mammals; development of advanced semihuman primates.
7				
			Miocene	Modernization of mammals; spread of grasslands.
25				
			Oligocene	Transition from more primitive to more modern mammals; first higher primates.
37				
			Eocene	Continued mammal expansion; beginnings of several modern groups.
55				
			Paleocene	Mammals expand rapidly on land, producing a large variety of archaic groups; marine fauna essentially modern.
65				
	MESOZOIC	Cretaceous		Extensive seas; first flowering plants, advanced sharks and bony fish; primitive mammals; dinosaurs, ammonoids, and other groups become extinct.
140				
		Jurassic		Dinosaurs and gymnosperms dominate land; molluscs, especially ammonoids, abundant in oceans; first birds.
195				

THE GEOLOGIC TIME SCALE

Age	Era	Period	Events
230	PALEOZOIC	Triassic	Molluscs dominate seas for first time; reptiles diversify; first dinosaurs and mammals; gymnosperms dominate plants.
280		Permian	Reptiles dominate terretrial environments; coal forests decline; massive extinction affects many groups.
310		Pennsylvanian	Origin of reptiles; climax of extensive coal forests.
350		Mississippian	Stalked echinoderms and brachiopods very abundant; amphibians diversify; forests expand
400		Devonian	Prolific sealife; first ammonoids; development of sharks, bony fish, and amphibians; first true forests.
440		Silurian	First large coral reefs; first terrestrial plants and animals; first jawed vertebrates.
500		Ordovician	Prolific sealife; first appearance of corals, bryozoans, crinoids and other echinoderms; expansion of brachiopods
570		Cambrian	Rise of shelled metazoans, especially trilobites and brachiopods; first vertebrates.
	PRECAMBRIAN		First singlecelled and multicelled animals.
			Advanced algae
4,700			Bacteria, blue-green algae, and stromatolites Origin of earth

BASIC PRINCIPLES

us to understand precisely where it fits in the progression of life, and makes it possible to compare our assemblage with all other known fossils of the same time, in order to obtain a broader view of the earth's inhabitants at any given time in the past.

Uses of Fossils

The study of fossils contributes many different kinds of information to our knowledge of the history of the earth and the organisms that have inhabited it. Fossils are the only record we have of the amazing variety of extinct organisms that have existed, then vanished, over the last 3.5 billion years. This record is a very rich one—several hundred thousand different types of fossil plants and animals have been described by paleontologists. It allows us to reconstruct the nature and progression of life from the far past to our own time; to visualize, albeit incompletely, the organisms and communities of organisms present along a 100-million-year-old seashore, or in a 250 -million-year-old forest, or around a 400-million-year-old coral reef. Fossils extend our perspective from the tiny slice of geologic time known as the present far back through the billions of yesterdays that are closed to mankind's direct scrutiny. Recognition and understanding of the changes in past life through the ages contributes to our understanding of how the organisms we see around us came to be, and ultimately, to an understanding of how our own species developed.

As living entities, humans share many basic life processes with other organisms. We have five-fingered hands with opposable thumbs—an inheritance from small arboreal creatures that lived 50 million years ago. We have hair, constant elevated body temperatures, and give birth directly to smaller replicas of ourselves—features first developed by the earliest mammals 200 million years ago. We breathe air with enclosed lungs, an invention of some early fish 400 million years ago. And many of our metabolic processes, as well as our male/female nature, are shared with primitive single-celled creatures that first developed them more than a billion years ago. The history of life, documented by the fossil record, is, in a very real sense, *our* history.

The underlying similarities of all life, past and present, demonstrate that each type of organism is genetically related to others, rather than being separately created. The fossil record documents, sometimes in great detail, progressive changes through time, from one kind of organism to another—the process known as evolution. We can observe in many sequences of sedimentary rocks the gradual change in the structure and morphology from one species to another, sometimes through entire lineages of species over millions of years. Because the fossil record is incomplete, we will always be ignorant of the details of some of these evolutionary transitions, but there is still an abundance of evidence that evolutionary change is a basic feature of life. One of the most significant aspects of the fossil record, then, is the documentation of these changes through the vast expanse of geologic time, which, when added to knowledge of the mechanisms by which these changes occur (derived mainly from biological studies of modern organisms), has established evolution as the most compelling explanation for the development and diversification of life.

As discussed in the previous section, the fossil record is the basis of the geologic time scale and the dating of sedimentary rocks. Because each fossil species developed from an ancestor, was successful, and then vanished, either by extinction or by evolving into something else, each fossil species represents a discrete, usually very short episode in geologic time. Recognition of the same species in rocks that may be widely separated allows geologists to correlate these rocks, to say that they were deposited during about the same relatively brief time. Establishing the exact order of these episodes (the succession of fossils) allows an extremely precise determination of the history of the earth, a major activity of biostratigraphers, and, indeed, a major goal of the science of geology.

Fossils also provide information about past environments. Reasoning primarily by analogy with the relationships of modern organisms with various modern environments, it is often possible to interpret paleoenvironments very accurately. The presence of fossil animals such as crinoids and brachiopods, found today only in the ocean, in Pennsylvanian rocks at the top of the Sandia Mountains, tells us that these rocks were deposited in a marine environment, even though the nearest sea is now 800 kilometers away. Similarly, the presence of ferns and large numbers of deciduous tree leaves, along with the scales and teeth of fossil fish, in Cretaceous rocks of northwestern New Mexico, strongly suggests a well-watered environment of dense forests and meandering streams—quite different from the environment of this area today. The study of fossils in an environmental context brings home with great force the fact that New Mexico has changed considerably through time, that seas have advanced and retreated across its face, that great swamps once existed where deserts are now, that mountains have been raised and then eroded to plains.

A concept of perhaps even greater magnitude than the impermanence of landscapes and environments, is that of plate tectonics. In what has been described as a revolution in the earth sciences, the surface of the earth is now understood to be composed of numerous rigid plates, moving relative to each other on an underlying semiplastic medium. As the continents are parts of these plates, they move too (continental drift). Paleontology provided some of the earliest evidence for continental drift, and continued analyses of the distribution of past organisms has aided in our understanding of the previous positions of continents. The presence, for example, of several types of late Paleozoic fossil land plants and animals on continents now widely separated by oceans, such as South America, Africa, and Antarctica, provides evidence that these continents were once sutured together into a single land mass. The fact that some marine Silurian fossils in New England and northeastern Canada are more closely related to those of England than to those found farther west in the United States suggests that North America and northern Europe were once much closer than they are today. An increasingly important emphasis in paleontology is the analysis of the ancient geographic distributions of organisms, or paleobiogeography.

Fossils have economic uses, too. Aside from the fact that coal and petroleum *are* fossils, an understanding of how and where they form is important to the prediction of where they might be found. Petroleum tends to accumulate in porous rocks, and much petroleum is found in the interstices of ancient reefs, large structures that were formed by ancient organisms in warm shallow seas. Identifying reef environments in the subsurface on the basis of fossils may lead to the finding of new oil reservoirs. Fossil logs in New Mexico Jurassic sandstones are part of an ancient stream environment in which uranium ores developed; in fact, it is possible that the organic material in the wood actually influenced the deposition of the uranium. As objects that allow the correlation of rock units, fossils may be extremely helpful in tracing ore-bearing formations in areas that have been extensively faulted or otherwise geologically disrupted.

Finally, fossils have appealed to people through history as interesting and sometimes aesthetically pleasing things. The mother-of-pearl layers of extinct ammonoids flash the colors of the rainbow, and petrified wood has long been polished and used in jewelry. Fossils are sometimes found replaced by opal, and have thus become this semiprecious gemstone. The heavy golden pyritized fossils found in a few places in the United States have been avidly collected from the time of their first discovery. A well-preserved trilobite is beautiful in the proportions and symmetry of its shape and structure, and stimulates a sense of wonder and admiration even among professional paleontologists. Fossils are as close as we will ever get to seeing living representatives of these long-extinct creatures, so different from anything living today. Fossils are a way of touching a fragmentary piece of the deep past— a past that was filled with animals and plants every bit as varied and successful as those we now live with.

Collecting Fossils

Preparation

As with any endeavor, collecting fossils requires a certain amount of advance preparation. Because fossils are present all over New Mexico, it is possible to drive to a productive site and collect from it within a day or less, no matter where in the state one begins the trip. Maps are essential, especially if travel off paved roads is anticipated. Topographic quadrangle maps, published by the United States Geological Survey, show topographic and geographic features very accurately, down to the locations of even minor dirt roads and isolated buildings. Seven and a half minute (1:24,000) quadrangles are available for most parts of the state; the rest are covered by 15 minute (1:62,500) quads. Geologic maps are also highly useful, because they show, usually in colors, the different rock units exposed in an area, allowing one to determine at a glance their names, ages, and lithologies. Because some formations are fossiliferous almost wherever they are exposed, whereas others yield few fossils, reference to a geologic map can be very helpful in deciding where to concentrate one's collecting efforts in the field. Unfortunately, detailed geologic maps are available for only a relatively small percentage of New Mexico quadrangles, but the number is increasing at a brisk rate as geologists study and produce maps for additional areas. A large geologic map of the entire state is available, and is useful in understanding the large-scale distribution of rocks of various ages across New Mexico. The maps mentioned above may be obtained at technical bookstores, and at the New Mexico Bureau of Mines and Min-

BASIC PRINCIPLES

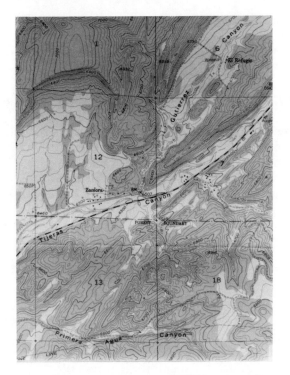

Part of a topographic map. The contour lines show elevation above sea level; many other topographic and geographic features are also portrayed.

eral Resources in Socorro, New Mexico. United States Geological Survey maps may also be ordered from the regional Survey office in Denver. (Branch of Distribution, U. S. Geological Survey, P.O. Box 25286, Denver Federal Center, Denver, Colorado, 80225)

In addition to maps, much information on the geology and paleontology of specific areas in New Mexico is contained in the published technical literature. Some useful works on New Mexico paleontology are listed in Part IV of this book, along with comments on where to go for further information.

While it is good to have some prior knowledge of a locality and the kinds of rocks and fossils likely to be found there before going into the field, it is sometimes rewarding to search areas where no fossils have been previously reported. New Mexico is a large state and by no means have all the fossiliferous strata and sites been located. One strategy that paleontologists use is to choose a

COLLECTING FOSSILS

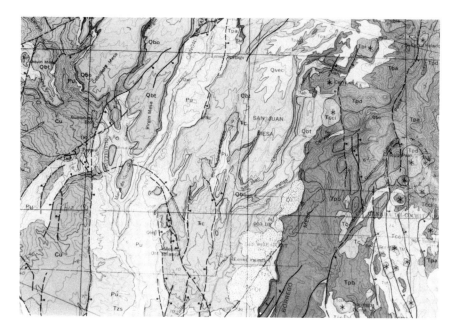

Part of a geologic map. Formations are portrayed in different colors on the map, along with other geologic features like faults (continuous and dashed lines). The geologic information is often superimposed on a topographic map, as illustrated here.

formation that is generally fossiliferous, and visit exposures of that formation that they have not been to before. This often leads to new collections and localities, adding to our information on the distribution of fossils within the state.

Before starting out, assuming a definite location is in mind, it would be wise to consult land ownership maps, available in United States Bureau of Land Management offices and in technical bookstores. These maps indicate who has jurisdiction over a given section of land, whether it is privately owned, or within Indian, state or federally managed areas. Access for the purpose of collecting fossils varies depending on the land's ownership. On privately owned land, the fossils belong to the owner of the land, and it is appropriate and courteous to obtain permission before collecting. Collecting on Indian reservations and Pueblos requires written permission from the tribal governor or council beforehand.

BASIC PRINCIPLES

Lands under state (12 percent) and federal (34 percent) jurisdiction together constitute nearly half of New Mexico's area. Fossil collecting is not regulated on state land at present, except in special areas such as state parks. Federal land includes national forests, parks, and monuments, as well as areas managed by the Bureau of Land Management. The philosophy of the federal government is that federal land, and resources such as fossils on or in the land, belong to all citizens of the United States. Thus, collection of fossils by individuals is regulated on some federal lands. Permission is required in order to collect from national forests, parks, and monuments. Regulations affecting fossil collecting on BLM land are in a state of flux at present. Permits are necessary in order to collect vertebrate fossils, but collection of plant and invertebrate fossils is not specifically discussed in existing federal regulations. Generally speaking, there are no restrictions on collecting from roadcuts along state and federal highways.

Equipment

Fossil collecting demands very little specialized equipment. One should dress appropriately to the terrain and climate; hiking or work boots are desirable in rough terrain. Intact fossil shells eroding out of the rock can be readily picked up by hand and put into collecting bags, which need be no more than a cellophane or paper bag for casual collecting. A knapsack or pack should be taken along to carry the bags of fossils that are collected. Geologists use rock hammers when specimens must be chipped out of the rocks; the best kind for sedimentary rocks has a head bearing a thick blunt surface on one end and a compressed flat surface on the other. The flat end allows easy insertion along bedding planes and facilitates the raising of thin slabs of shale or sandstone. Chisels are helpful in this regard too. Plastic containers with caps are best for storing small fossils; if they are fragile, wrap them in several layers of tissue paper before putting them into the container. Fossils that are large but fragile should be wrapped initially in tissue paper and then in newspaper; this helps prevent the specimen from crumbling, and even if it does, the pieces will stay more or less together. Masking tape can be used to secure the newspaper wrapping. Some types of fossils, such as carbonized plants, will dry upon exposure to air and tend to flake off in a very short time. Spraying on a thin coat of transparent shellac, such as Krylon, inhibits flaking and provides a protective coating for these delicate fossils. Glue or Duco Cement is handy for on-the-spot repairs

COLLECTING FOSSILS

Collecting equipment. These are the things I usually take out into the field to collect invertebrate fossils. If large slabs of material are to be collected, short-sided cardboard boxes are also useful as containers for the slabs or other large fossils.

of fossils broken during collecting and watered-down glue or shellac may be used to harden brittle fossils. A hand lens (10-power) is useful for observing the details of very small fossils, and having one in the field often helps in the immediate identification of a small specimen.

Needless to say, the types of equipment used by a fossil collector vary with individual preference and experience, and with the type of fossils at a given locality. Paleontologists occasionally have need of more elaborate and exotic equipment, ranging from materials for applying plaster jackets around vertebrate bones; to burlap sacks and screening boxes for processing and concentrating nearly microscopic mammal teeth from bulk sediment collections; to bulldozers, backhoes, rock saws, and jackhammers for really large excavations.

Two pieces of equipment are absolutely essential for every fossil collector: a notebook and a pen or pencil. One should get into the habit of recording as much useful information as possible about the location and rock formation from which fossils have been collected. In addition to keeping a notebook, locality and stratigraphic information should be written on a slip of paper and put with each specimen or collection, so there will never be any doubt about where the specimens came from. Do not trust to memory! If the appropriate United States Geological Survey topographic map is available, localities can usually be pinpointed with great accuracy. This is important, particularly if rare or unusual specimens are found, because additional specimens can be collected later from the same site. Each fossil can be thought of as a package of information about ancient life, but much of this information can only be retrieved if we know (or can find out by going back to the site) its exact stratigraphic position, the type of rock it came from, its orientation within the rock, and its relationships to other fossils in the same sedimentary unit. A fossil without accurate locality information may be a pretty object, but it is almost valueless in a scientific sense.

Ethics of Collecting

People collect fossils for many different reasons, but all collecting removes fossils from their stratigraphic context and diminishes, to a greater or lesser degree, the amount of information that can be obtained from them. In general, the scientific study of fossils is not adversely affected by laymen collecting from deposits that have previously been thoroughly studied by paleontologists. Paleontologists do worry about collections that are made of rare or unstudied fossils, because in some cases this removes the specimens from scientific scrutiny, and the new information they may provide about the history of life may be lost to professional paleontologists and interested amateurs alike. On the other hand, nonpaleontologists have been instrumental in the finding of many new and interesting types of fossils, and by their collecting activities have considerably enriched our knowledge of ancient life forms. These points are particularly relevant to New Mexico because this state has vast numbers of fossils, but relatively few of its fossil deposits have been well studied. The potential for exciting new discoveries being made by paleontologists as well as amateurs is great.

In this context, cooperation between paleontologists and amateur fossil collectors is essential. This should be easy because both groups are driven by similar motivations—the love of discovery, the enjoyment of finding and learning about these remnants of ancient life, an appreciation of the beauty in their shape and form, a desire to understand the diversity of past life, the need to be doing things out-of-doors away from the congestion of city life—the list could go on and on. Because fossils are ultimately a "nonrenewable" resource, it is appropriate that all fossil collectors reflect upon the meaning and ramifications of their collecting.

A fossil-bearing exposure is part of the larger environment of an area, and it is important that this is not adversely affected during the collection of fossils. Do not carry anything into an area that you do not carry out again, and use off-road vehicles with great care; fossils as well as living plants and animals may be harmed by being run over. There is never a need to collect every last fossil from a site. Remember that others coming after you are entitled to the enjoyment you experienced while collecting, and that only a few specimens of each kind from a given site are needed for a representative collection.

Any excavations that are made in the course of collecting should be filled in and the site returned to its natural state before departing. As fossils erode readily from most kinds of sedimentary rocks, it will rarely be necessary to excavate extensively in order to collect them. Large bones or other fossils requiring specialized techniques to remove are best left where they are; trying to collect these may severely damage or even destroy the specimen. In such cases, a good photograph of the intact specimen will be more meaningful than the fragments that might result from trying to remove it. A paleontologist can advise on the value of the find and whether the specimen stands a reasonable chance of being extracted with minimal damage. In recent years there have been several notable finds of large vertebrate fossils in New Mexico by amateurs, who then participated with professional paleontologists in the excavation and preparation of the specimens.

The ultimate fate of the fossils one collects is important to consider. Putting specimens in an organized collection with good locality records, with the intention of learning more about them and perhaps exhibiting them is quite a different matter from a casual collection picked up, briefly looked at, and then tossed out the car window on the drive home. It would be preferable to refrain from collecting, and just examine the fossils in the field, rather than collecting merely to discard the specimens soon afterwards.

It is inevitable that a collector will occasionally find a fossil that cannot be identified. If such is the case, or if more information is desired about certain fossils, there are several paleontologists in New Mexico who will be happy to examine your find and provide more information about it (see Part IV). If a specimen is new to science or otherwise of great interest, the paleontologist may ask to borrow it for study. The study and subsequent publication of new information on a fossil is the main way in which your discovery is made known to scientists and all others who are interested in paleontology.

Specimen Preparation

Although some fossils weather cleanly out of the rock and need no preparation to show their characteristic or interesting features, many specimens require additional attention after they have been collected. The most common problem is one of matrix adhering to the fossil; depending on the type of matrix, several ways of cleaning fossils are available. If the matrix is relatively soft shale or sandstone, it can often be flaked off with a sharp-pointed probe. Dental supply houses have a wide variety of instruments that can be used for this purpose. Judicious use of a small sharp chisel may also help, in cases where the fossil is relatively large and unfragmented. Where the matrix is a hard limestone, it is sometimes impossible to cleanly extract a specimen; as a last resort, hitting the area around the fossil may cause it to pop out of the matrix with little damage, but one always runs the risk of cracking the specimen. Fossils with a little limestone adhering can sometimes be cleaned up with a dilute (5 percent) solution of hydrochloric or acetic acid, which readily dissolves limestone. When silicified fossils are present in a limy matrix, the entire piece can be dropped into acid, which will entirely dissolve the limestone without affecting the silica that composes the specimen. This technique, acid etching, has been applied on a large scale by paleontologists to fossils of the Permian reef complex of southeastern New Mexico and West Texas, and many tons of limestone have been processed in this way, yielding thousands of delicately preserved silicified invertebrates that could never have been otherwise extracted. Acid should be used with great care, always highly diluted, and in places with adequate ventilation, for the acidic vapors may burn the lining of the lungs.

A very effective (but very expensive) instrument in wide use in paleontological laboratories is the airbrasive unit, or air dent. This machine, using compressed air, sends a high velocity stream

SPECIMEN PREPARATION

of powder particles through a nozzle and "sandblasts" the matrix from around a fossil. This works well only when the fossil is harder than the matrix. Careless use of the airbrasive results in the erosion of the specimen, and the creation of "new species" by the wearing away of certain structures or the reshaping of the form of the specimen. More affordable to the average fossil collector are small electric hand-tools with vibrating points that act as a miniature jackhammer in flaking matrix off a specimen.

Fossils collected in bulk samples, as with small specimens present in great numbers in samples from a shale bed, may be released from the matrix by chemically disaggregating the shale. A first step is to put the sample in a pan and soak it in water for a few days; then let it completely dry. If this is not sufficient, soaking in kerosene or a strong detergent may facilitate the disaggregation of matrix without harming the fossils.

Fragile specimens preserved on bedding planes, such as carbonized plants, sometimes require very careful preparation using pointed probes, to scrape away excess material that partially covers the fossil. Preparation in this manner often reveals much more of the fossil than was visible when it was collected, and makes for a more complete and cleaner specimen. Glue, or better, Duco Cement, should be available for reuniting pieces of a specimen that have become detached during collection or preparation. For fossils in pieces too large to glue together, like fragments of a dinosaur bone, a material called water-putty is excellent. It is a powder that becomes sticky and moldable when water is added, and can easily be applied to the broken edge of two fragments to stick them together; once in place it soon forms a very hard connection between the joined pieces. Various epoxy solutions may be used as well.

As most readers are already aware, some types of fossils (like petrified wood) can be cut and polished with lapidary equipment to emphasize the color and texture of the specimen. A detailed consideration of these techniques is beyond the scope of this book. I only point out here that while cutting and polishing may contribute greatly to enhancement of the decorative aspects of a fossil, they should not be attempted on fossils that are rare, unusual, or of scientific interest. These types of fossils should remain as intact as possible.

Each collector will determine the best way to display his or her fossils, in accord with individual preferences and purposes. I offer only one suggestion based on my own experience—that fos-

BASIC PRINCIPLES

sils may be best appreciated if they can be handled, their texture and shape felt, and their features examined closely from all angles. Only the most fragile of fossils need be removed from physical contact with interested observers. Needless to say, no fossil should be displayed in such a way that part of the fossil is altered or damaged.

2

THE MAIN FOSSIL GROUPS

The fossil record encompasses an incredibly diverse set of organisms. Out of the 100,000 to 200,000 different kinds that have been described, more than 5,000 different types of fossils have been reported from New Mexico rocks. In order to understand the New Mexico fossils that will be discussed later, we first must have some appreciation of the general features of living things, and of the major groups of organisms that have developed and diversified through time. Because fossils represent ancient organisms, they are included in the same system of classification used for all living organisms. Though almost all distinct kinds (or species) of fossil animals and plants are extinct, most of the major groups to which ancient species belong have endured for hundreds of millions of years, and we have living representatives existing today to observe. The purpose of this part of the book is to introduce the way in which fossils are classified, and to explore the general characteristics of the major groups of organisms that have left a fossil record in New Mexico rocks.

Classification

There are about 1.5 million species of plants and animals living today, not to mention the more than 100,000 known fossil species, and hundreds of new fossil plants and animals that are described every year. To make sense out of this staggering diversity, and to understand better the relationships between the various kinds of organisms, past and present, biologists and paleontologists have long used a system of classification developed in the mid-1700s by a Swedish botanist, Carl von Linné (the

THE MAIN FOSSIL GROUPS

name is often Latinized to Linnaeus). Although many criteria could have been used (and had been used earlier) to construct a classification system, Linné's great contribution was to choose similarity of form and structure as being the most natural and fundamental features on which the true relationships between living organisms are based.

In the Linnaean system of classification, each kind of organism is assigned to a hierarchical series of groups. Types of organisms (species) that closely resemble each other are considered to belong to a group called a genus; similar genera are grouped into a family; similar families into an order; similar orders into a class; similar classes into a phylum; and similar phyla into a kingdom. Each group higher in the hierarchy is more general and includes a wider variety of organisms than the one below it. The highest level, the kingdom, is the most general group of all, including, for example, all animals or all plants.

To indicate how this works in practice, the classification of the house cat is given below:

KINGDOM: Animalia — All animals.
PHYLUM: Chordata — All animals with a notochord and gill slits (or derivatives), most also having a skeleton of bone or cartilage (subphylum Vertebrata), such as mammals, reptiles, amphibians, sharks, bony fish, and so on.
CLASS: Mammalia — All warm-blooded hairy chordates with a single bone in the lower jaw, such as carnivores, rodents, elephants, whales, even-toed ungulates, bats, and so on.
ORDER: Carnivora — All mammals with carnivorous teeth and skeletal features, such as the cat, dog, bear, raccoon, and hyena families.
FAMILY: Felidae — All true cats: domestic cats, lions, tigers, cheetahs, and so on.
GENUS: *Felis* — A type of true cat, including the domestic cat, mountain lion, bobcat, and so on.
SPECIES: *catus* — The domestic cat.

As can be seen, there are seven main levels to the hierarchy, but a large number of other categories (e.g., subfamily, superfamily, and so forth) are available if needed. Each species of living and extinct organism belongs to the Linnaean hierarchy and each has its unique place in it. By examining such classifications, we can immediately appreciate the relationship of any species with any other species. For example, we can tell that domestic cats are closely related to mountain lions (*Felis concolor*) because the two species are in the same genus. House cats are only moderately closely related to dogs, however (they are in the same order, but share no major lower category), and house cats are only remotely related to insects (they are in the same kingdom—both are animals—but that is all).

An important part of classification is nomenclature, the application of unique names to each kind of organism. The scientific name of a species is a binomial term consisting of the genus and species names; hence, *Felis catus* is the scientific name for the house cat. Scientific names are usually constructed from Latin or Greek words, but may also come from the name of a person or a place; for example the Pennsylvanian gastropod *Taosia* is named after Taos, New Mexico, the place where it was first discovered. More than a dozen fossil species are named after the state of New Mexico. Because each species has a unique name, which may not be applied to any other organism, past or present, when we use the name (no matter what our native language might be) there is no question as to exactly what organism we are referring to. When a new type of fossil or living organism is found, and hundreds are discovered every year, it is immediately described thoroughly, named, and fitted into the Linnaean system of classification. It is thus uniquely defined. One of the reasons the Linnaean system has worked so well over the 200 + years since its introduction is because it reflects, as well as any man-made classification scheme can, the evolutionary and genetic relationships between organisms, as expressed by all of the details of their form and structure.

The Main Divisions of Life

Traditionally, life has been divided into two kingdoms—plants and animals—separated by very basic differences in construction and operation. Plants are green (have chlorophyll), photosynthesize (make food from water, CO_2, and sunlight, giving off oxygen as a waste product), have thick walls around their cells, and generally lack muscular, sensory, nervous, or locomotory structures; whereas animals use other organisms for food (either plants or other animals), take in oxygen, and give off CO_2, lack chlorophyll, have only membranes around their cells, and generally possess in one form or another the structures mentioned above. Though

THE MAIN FOSSIL GROUPS

recent studies have stressed even more fundamental divisions of life, as between prokaryotes (bacteria and primitive algae) and eukaryotes (advanced algae and everything else), and have also added at least two new kingdoms, Monera (prokaryotic organisms) and Protista (all single-celled organisms whether or not they function as animals or plants), the kingdoms of the plants and animals are still the most easily recognized and frequently encountered major divisions of life.

Most fossil protistans that have left a good record functioned as animals, and for convenience will be included under animals in this book. Protistans, or protozoans, however, differ from animals (or metazoans) in being composed of only a single cell, whereas metazoans are multicellular. Thus protozoans are usually microscopic, and must carry out all the life functions of the organism, often including the secretion of a skeleton, within a single cell. Multicellular animals, on the other hand, are often large, have billions of cells, and have these cells segregated into specific structures, like organs and tissues, that are responsible for a single restricted function in the animal's life processes. The transition from protozoan to metazoan is undocumented in the fossil record, occurring in the late Precambrian before organisms had developed the ability to secrete fossilizable hard parts, but it stands as one of the most important events in the history of life.

Finally, the terms invertebrate and vertebrate are often used as major informal divisions of the animal kingdom. Invertebrates include animals lacking a skeleton composed of bone; vertebrates are animals that have a bony or cartilaginous skeleton. This division of animals is a very unequal one. About 95 percent of animals, past and present, have been invertebrates; only part of a single phylum, Chordata, are vertebrates. Vertebrates have been of interest to a degree far outweighing their relatively small numbers because humans are vertebrates, and so are the majority of animals we are most familiar with through domestication, as food sources, and so on. In addition, vertebrates include the most complex organisms that have developed on earth (and perhaps anywhere in the universe); some vertebrates have prodigious memories, a desire to learn even when the knowledge is not necessary to solve an immediate problem, and the capacity for rational thought; they construct elaborate cultures that have superseded purely biological factors in controlling their subsequent evolution.

Within the animal and plant kingdoms are numerous major subdivisions called phyla. These represent important distinctive

ANIMALS AND ANIMAL FOSSILS

body plans that are usually easily recognizable; identifying a fossil to its phylum is the first step in its precise identification, and in the understanding of how that fossil lived and functioned. Only about thirty animal phyla have ever existed; only one phylum (the Cambrian archaeocyathids) that we know of has become extinct. Several animal phyla are known only from bizarre and uncommon creatures living today in the ocean, and only about a dozen phyla have left a decent fossil record. These are introduced a few pages farther on.

Animals and Animal Fossils—Some General Features

The shape and structure of an organism and its fossilized skeleton reveal more than merely its name and relationships. Each organism is a package of adaptations that allow it to function successfully in a particular environment. By understanding some basic features of animals we can obtain much information from, for example, a fossil shell, that indicates how the animal functioned when it was alive, and how it went about its "day-to-day activities."

Two aspects of animals that strongly influence shape and structure are mode of life and feeding methods. There are three fundamental modes of life in the water—swimming (nektonic), floating (planktonic), and substrate bound (benthonic). Benthonic organisms include those that crawl on, burrow into, or are attached to the sediments. Obviously, an animal that swims will have structures appropriate to that mode of life, and these will be different from the equivalent structures that a stationary attached benthonic animal will have. Development of the same mode of life in unrelated animals often produces superficially similar organisms (convergent evolution); for example, sharks and dolphins look similar because there are only a relatively few types of adaptations that are optimum for swimmers—fins, widened tail, streamlined shape, and so on. On the other side of the coin, one reason for the great diversity of organisms within some phyla is that members of the phylum have evolved a wide variety of modes of life. Within the phylum Mollusca, for example, clams burrow, snails crawl around on the sediment surface, and octopuses swim. Except for the fundamental features that make all of these animals molluscs, the shape and structures are very different, but always appropriate to a specific mode of life.

Likewise, there are only a limited number of ways to obtain

THE MAIN FOSSIL GROUPS

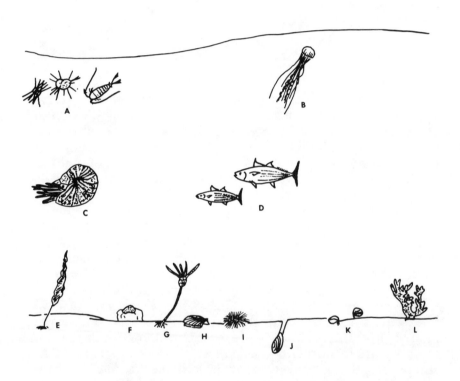

Modes of life in the ocean. **Planktonic (floating) organisms:** A, diatom, foraminifer, copepod (small arthropod); B, jellyfish. **Nektonic (swimming) organisms:** C, coiled nautiloid; D, bony fish. **Benthonic (substrate bound) organisms:** E, algae (seaweed); F, bryozoan, encrusting a rock; G, crinoid, attached by holdfast; H, pectinid pelecypod (lies free on sediment, occasionally swims); I, echinoid (moves slowly across sediment); J, pelecypod (burrows into substrate); K, brachiopod, attached by a fleshy pedicle; L, coral, attached to substrate. *(Modified from McAlester)*

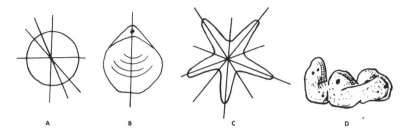

Types of symmetry. A, radial; B, bilateral; C, pentameral; D, asymmetrical.

food. Many animals obtain other organisms to eat in a variety of ways, ranging from predaceous carnivores that chase and capture other animals, to snails that scrape algae off an intertidal rock. A large number of marine invertebrates, particularly those that are attached to the substrate, have structures for filtering microscopic organisms or bits of organic matter out of the water around them. Some marine animals can actually absorb dissolved nutrients from the water. Other animals eat their way through the sediment, absorbing the organic material attached to the grains and eventually expelling the sediment from their digestive tracts; earthworms are a familiar example of this deposit-feeding behavior. Parasitism is a less common feeding method and evidence of parasitism is rarely directly seen in fossils.

In addition to mode of life and feeding methods, the symmetry of an organism places constraints on its shape and the structures it might develop. Symmetry is a fundamental aspect of the form of an organism, and can most easily be determined by imagining a plane that perfectly bisects an organism into left and right mirror images. If there is an infinite number of ways this can be done (as with a plane through the center of a circle), the symmetry is said to be radial. If there is only one way this can be done (imagine a plane bisecting your body into left and right halves), the animal has bilateral symmetry. Animals with bilateral symmetry are the only ones with portions of the body that can be designated as front and back, and it is no accident that bilaterally symmetrical animals have their sensory and perceptual structures concentrated into the part of the body that meets the environment first—the head. Pentamerous symmetry is seen in only one phylum, the echinoderms; a good example is the seastar. As the

THE MAIN FOSSIL GROUPS

Solitary and colonial organization. At left is a solitary coral—one individual animal and its skeleton. At right is a colonial coral, in which several individuals are physically attached, sharing connections in some soft part structures as well as in their external skeletal elements. Most groups of animals, including humans, are solitary.

name implies, there are five different positions a plane of symmetry may have to divide the animal into left and right mirror images. Animals with a form that cannot be divided into left and right mirror images are asymmetrical. A knowledge of symmetry is important not only because symmetry strongly influences an animal's shape, but also because some animals that look superficially similar (such as clams and brachiopods) are oriented differently with respect to a plane of symmetry, and this reflects basic differences between them and helps in their identification.

Most animals are solitary—each individual exists by itself without a physical connection to other members of the species. Humans and almost all familiar animals are solitary. A few phyla, however, include individuals that are attached to neighboring individuals, often sharing the same digestive tract or other internal structure, and forming a colony of hundreds or even thousands of connected animals. Corals and bryozoans are the two main groups having colonial species, and these can often be recognized by the presence of a colonial skeleton with numerous compartments where the individual organisms lived.

Still another feature shared by most animals, but in many different variations, is the presence of a skeleton that supports the body as well as individual internal and external soft structures. Most invertebrates have exoskeletons that surround or cover the soft parts (a clam shell, for example), but some groups, like vertebrates, have an internal skeleton, or endoskeleton. Though most skeletons, particularly those that have become fossilized, are rigid and mineralized, soft-bodied organisms may have skeletons of muscle, internal fluid-filled cavities, or coverings composed of resistant organic molecules. Arthropod exoskeletons, for exam-

INVERTEBRATE FOSSILS

ple, are thin, flexible, and composed of the organic material called chitin. Mineralized skeletons, because of their composition and durability, are far more likely to become fossilized than non-mineralized skeletons, and this adds an element of bias to the fossil record. Insects were probably the most abundant and diverse group of animals 100 million years ago, just as they are today, but the fossil record does not reflect this; instead, fossils of 100-million-old insects are rare, at least partly because they lack hard mineralized skeletons.

With these general points about the factors influencing the form and structure of organisms in mind, we now turn to a summary of the major groups of animals and plants that have left a good fossil record in New Mexico. Then, in Part III we will see how these groups are represented in New Mexico rocks of various geologic ages.

Invertebrate Fossils

Fossils of invertebrates, especially marine invertebrates, are by far the most common and widespread types of fossils found in New Mexico. The major groups of invertebrates are described briefly below, and illustrations of representative specimens are provided to show the important distinguishing features of each group.

Protozoans

Protozoans include a wide variety of single-celled organisms. Most are very small, and many are microscopic, but some may be 2 cm long or more, with amazingly intricate and complex skeletons for such small creatures. The most abundant protozoans in New Mexico rocks are foraminifers ("forams"). Through their long history forams have developed a tremendous range of shell shapes, and a fair number are large enough so that they can be spotted in rocks with the naked eye. Forams generally have a perforated calcareous shell consisting of one to many chambers; some are able to construct shells out of minute sand grains or other material. Until the Cretaceous Period, all forams lived on the sediments, mainly in the oceans, but at that time planktonic forms with inflated chambers developed, and these types are still one of the most abundant organisms in modern oceans, and are an important source of deep-sea sediments. Coiled forams like

THE MAIN FOSSIL GROUPS

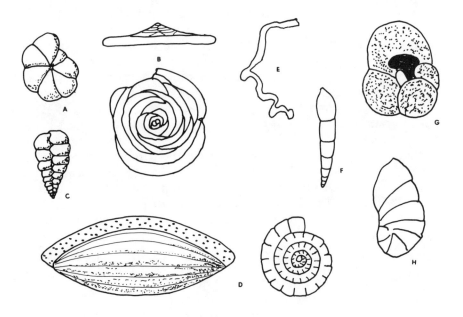

Protozoa. All are foraminifers, magnified 10 to 20 times; all are found in New Mexico. A, *Endothyra* (Mississippian); B, *Tetrataxis*, side and top views (Pennsylvanian); C, *Textularia* (Pennsylvanian); D, fusulinid, external view and cross-section (Pennsylvanian-Permian); E, *Apterinella*, often found encrusting the shells of other animals (Pennsylvanian); F, *Dentalina* (Cretaceous); G, *Globigerina*, a planktonic foram (Cretaceous); H, *Marginulia* (Cretaceous).

Endothyra are particularly characteristic of Mississippian rocks in New Mexico, and the relatively large (the size of a grain of rice) football-shaped fusulinids are abundant in the limestones of the Pennsylvanian and Permian Periods in the state. Many different kinds of microscopic forams, including planktonic ones, are found in the Cretaceous strata of northern New Mexico.

Porifera (Sponges)

Sponges are the most primitive multicellular animals. In life they are little more than large agglomerations of poorly integrated cells that spend their entire lives filtering water through a maze of canals and pores that perforate the animal. Sponges lack a head, brain, nervous system, true muscles, digestive tract, and other

INVERTEBRATE FOSSILS

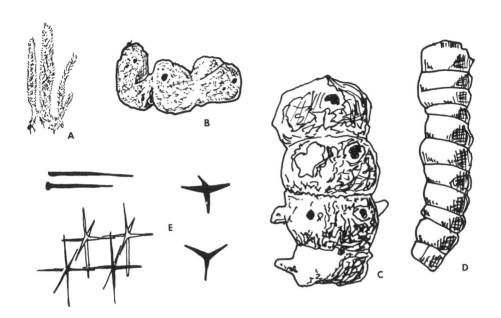

Porifera (sponges). A, B, two modern forms, × .5; C, *Girtyocoelia* (Permian), × 2; D, *Amblysiphonella* (Pennsylvanian-Permian), × 1; E, several kinds of isolated sponge spicules, to show diversity in shapes, × 5–10.

internal organs, and grow attached to the seafloor. Their skeleton is composed of numerous small to microscopic elements called spicules, which are composed of $CaCO_3$, SiO_2 (in the majority of sponges), or organic materials. Unless the spicules have fused to each other, in which case the general shape of the sponge is maintained after death, most sponge fossils consist of isolated spicules, or of an irregular pile of them, formed as the body of the sponge slowly collapsed as the soft parts were decaying. Modern and fossil sponges are asymmetric, growing in round, flat, or sometimes branching masses; many are totally amorphous. Few sponges have been reported from New Mexico rocks except for localized concentrations in strata of the great Permian reef system, in which sponges were an important constituent.

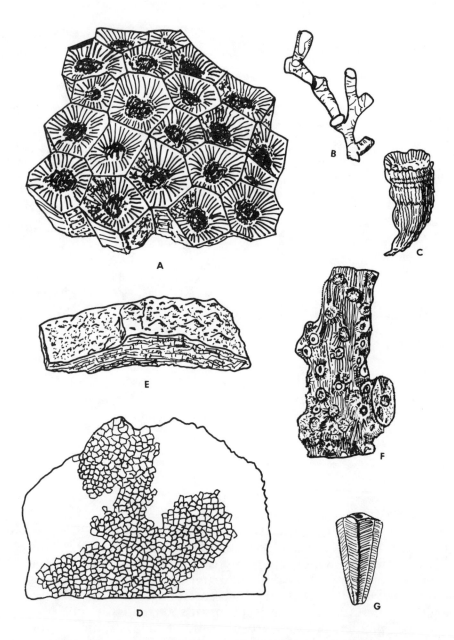

Coelenterata (all from New Mexico). A, *Hexagonaria*, a colonial rugose coral (Devonian), x 1.5; B, *Cladochonus*, a branching tabulate coral (Pennsylvanian), x 2; C, *Lophophyllidium*, a solitary rugose coral (Pennsylvanian), x .8; D, *Favosites*, a massive tabulate coral (Silurian), x 1; E, Stromatoporoid (Ordovician), x 1; F, *Archohelia*, a colonial scleractinian coral (Cretaceous), x 1; G, Conularid (Pennsylvanian), x 1.

Coelenterates

Coelenterates have two very different possible body plans: medusoid (jellyfish), and polypoid (corals). Ancient jellyfish are rarely preserved because they lack hard parts, so nearly all fossil coelenterates are corals of various kinds. The coral polyp is a small-to medium-sized, radially symmetrical, carnivorous animal, attached to the substrate and oriented with a mouth on the upper surface surrounded by tentacles filled with stinging cells. Some corals are solitary ("horn corals"), with a single polyp secreting a skeleton that is shaped like a curved cone and possessing many septa (thin vertical walls) radiating from the center to the periphery of the circular opening. The majority of fossil corals are colonial, with the massive or branched colonial skeleton having compartments for sometimes hundreds of polyps. These compartments are large enough to be easily seen, and in many corals are packed together to form a network of circles or polygons over the surface of the colony. Nearly all corals have calcareous skeletons.

Corals have been successful since their first appearance in the Ordovician, and at various times (like the present) have been important constituents of the large marine mound-shaped organic structures known as reefs. Three main groups of corals are found as fossils: (1)Tabulate corals—always colonial, generally with small polyp compartments and no radiating septa, and limited to Paleozoic rocks; (2)Rugose corals—may be colonial and sometimes very large, or solitary horn corals, always with radiating septa, often with large polyp compartments, also limited to the Paleozoic; (3)Scleractinians—modern corals, colonial or solitary, with septa, present only in Cretaceous rocks in New Mexico. Stromatoporoids, large layered and wrinkled fossils appearing as encrustations or round masses in New Mexico middle Paleozoic rocks, have an internal structure of tiny, often cubic compartments, and are considered by some authorities to be coelenterates, by others to be sponges.

New Mexico has good assemblages of Late Ordovician corals, and tabulate and rugose corals are locally common in Silurian through Permian rocks within the state, though in general they are relatively minor parts of the marine fauna. The great sea that covered much of New Mexico and the western United States during the Late Cretaceous was apparently not conducive to the growth of corals; Cretaceous corals are rare in New Mexico. An

THE MAIN FOSSIL GROUPS

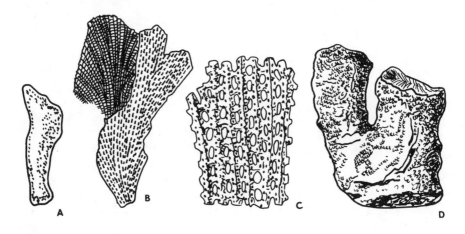

Bryozoa (all from New Mexico). A, Trepostome (Ordovician), x 1; B, *Fenestella* (Devonian-Permian), x 1; C, *Fenestella*, close-up, x 8; D, *Fistulipora* (Pennsylvanian-Permian), x 1.5.

unusual coral thicket was described a few years ago from Cretaceous strata near Lamy.

Conularids are small, pyramidal, presumably planktonic coelenterates with a thin phosphatic skeleton. They are rare, but distinctive, fossils in New Mexico Paleozoic rocks.

Bryozoa

Bryozoa, or "moss animals," are small colonial animals with a great variety of skeletal forms. Though superficially resembling small corals, bryozoan colonies are almost always smaller, and the individual bryozoan animals are nearly microscopic, for their compartments in the colonial skeleton appear as no more than pinpricks to the naked eye. They grew attached to the seafloor or to the shells of other animals. Bryozoans are one of the most

INVERTEBRATE FOSSILS

common types of fossils found in the marine Paleozoic rocks of New Mexico; they are essentially absent from later deposits within the state although the group is successful, but inconspicuous, in modern oceans. Most bryozoan fossils are broken pieces of branched colonies (they look a little like small twigs), fenestrate types with colonies that resemble a miniature trellis, thin encrustations on the shells of other animals, especially brachiopods, and small marble-sized masses.

Brachiopods

Brachiopods, though uncommon and inconspicuous in modern oceans, are the most abundant and diverse fossils found in Paleozoic marine sedimentary rocks, and brachiopods are conspicuous and important components of virtually all Paleozoic marine assemblages in New Mexico. These fossils consist of a pair of shells, or valves, that fit closely together, and this gives them a superficial resemblance to clams. Nearly all brachiopods are articulates, with their shells connected at a hingeline and articulated by tooth and socket structures on the inside edge of the hingeline. A few types are inarticulate—the two shells are held together only by muscles—and the composition of the shells is phosphatic rather than calcareous, with the result that fossil inarticulates are usually a dark purple or black color.

The two valves of a brachiopod are always slightly different in size and shape; the smaller valve represents the top side and the larger valve the bottom side of the living animal. A plane of symmetry dividing a brachiopod into left and right mirror images could only be oriented perpendicularly to the plane of the shells; in other words, it would divide both shells into left and right halves. This is in direct contrast to most pelecypods (clams), in which the plane of symmetry passes between the two shells and is parallel to them, making each entire shell a mirror image of the other.

Brachiopods are generally small animals, less than 2–3 cm wide, and show a tremendous variety of shell shapes. Most brachiopod shells are ornamented with fine to coarse radiating ridges and many have raised concentric wrinkles as well. The large highly convex forms called productoids often have fine spines over their shells that become detached during burial, leaving only tiny raised mounds on the shell. Many brachiopods have a circular or triangular hole at the back of the shells near the hingeline; a flexible rod (pedicle) protruded through it when the animal was

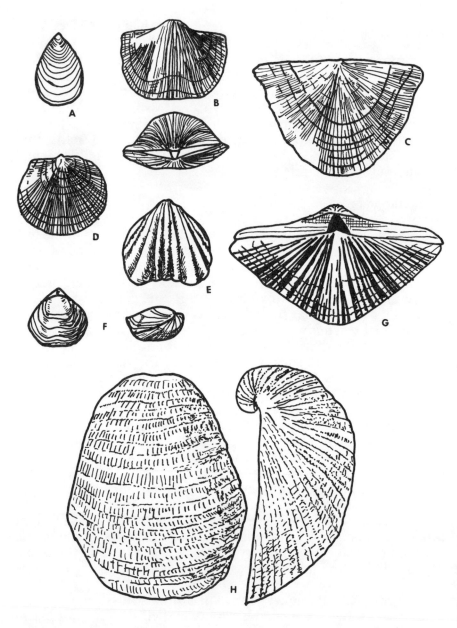

Brachiopoda (all from New Mexico). A, Inarticulate linguloid (Ordovician), x 2; B, *Hebertella*, top and back views (Ordovician), x 1; C, *Rafinesquina* (Ordovician), x 1; D, *Derbyia* (Pennsylvanian-Permian), x .75; E, *Lepidocyclus* (Ordovician), x 1; F, *Composita*, top and side views (Pennsylvanian-Permian), x .75; G, *Neospirifer* (Pennsylvanian-Permian), x 1; H, *Echinaria*, a large productoid, bottom and side views (Pennsylvanian), x 1.

INVERTEBRATE FOSSILS

Annelida. U-shaped burrow probably made by a large marine worm. The curved marks in the center of the U represent changes in the position of the burrow as the worm migrated up and down in the sediments (Cretaceous of New Mexico).

living, anchoring the shell to the substrate. When some erosion of a brachiopod shell has occurred, the intricate, often spiral internal structures that supported the filter-feeding tentacles may occasionally be seen.

Brachiopods are abundant and varied in New Mexico rocks from the Ordovician through the Permian. They are rare in Cambrian and Cretaceous marine deposits.

Annelids

The annelids, or segmented worms, are represented today by large numbers of terrestrial ("earthworms") and marine forms. Annelids have a very poor fossil record because most of them lack a mineralized skeleton. A few annelids, such as the spirorbid worms, have a tiny coiled calcareous shell and are sometimes found attached to the shells of other organisms. Because many annelids spent their time burrowing through sediments, trace fossils of worm burrows are moderately common in some New Mexico Paleozoic and Mesozoic nearshore marine sandstones and mudflat shales. While it would be difficult to prove that a given burrow was actually made by a worm, rather than some other kind of burrowing invertebrate, we know that some modern worms make burrows that are very similar to fossil burrows, and it is reasonable to surmise that annelids were responsible for at least some of the fossil burrows. Large, vertically oriented, U-shaped burrows found in some New Mexico Cretaceous strata, for example, were probably due to the activities of worms.

Molluscs

The phylum Mollusca is one of the most advanced and successful groups of invertebrates. It includes several distinctive body plans that resulted from the early divergence of molluscs into very different modes of life. Molluscs have well-developed muscles, an advanced nervous system, a wide variety of internal organs, gills, and one or more calcareous shells; and most molluscan groups have heads with brains and excellent sensory capabilities. The most familiar molluscan classes are gastropods (snails), pelecypods (clams), and cephalopods (squids and octopuses). These are considered separately below.

Gastropods. Gastropods are characterized by a single unchambered shell. In most gastropods the shell is coiled spirally, with the soft parts of the animal present throughout the inside of the shell, but the earliest forms had shells coiled in one plane and some later gastropods have reduced the shell to a cap-shaped structure or have even lost it entirely. Most of the body of a gastropod occupies the last largest whorl, and most gastropods crawl slowly over the substrate on a large muscular foot, feeding on plants, a mode of life exemplified by the modern garden snail. Although most gastropods are (and were) marine, several groups have colonized freshwater environments, and some have become fully terrestrial. In New Mexico, gastropods are subsidiary members of marine faunas, especially in the Ordovician and Mississippian-Permian, but they become extremely abundant in Cretaceous marine sediments. A few are present in Cretaceous and early Tertiary freshwater sediments in northwestern New Mexico; many

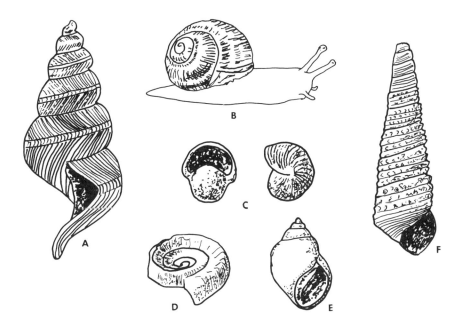

Gastropoda (all from New Mexico). A, *Hormotoma* (Ordovician), x 2; B, modern garden snail, showing foot and head, x 1; C, *Bellerophon*, front and side views (Mississippian-Permian), x 1; D, *Amphiscapha* (Pennsylvanian-Permian), x 1; E, *Lunatia* (Cretaceous), x 1; F, *Turritella* (Cretaceous), x 1.

terrestrial and freshwater forms are known from the Pliocene and Pleistocene of eastern New Mexico.

Pelecypods. Pelecypods possess two shells connected at a hinge-line by a tooth and socket articulation; they are opened and closed by a pair of large muscles connecting the internal surfaces of the shells. The two shells make up the left and right sides of the animal and in most pelecypods (oysters are the main exception) are mirror images of each other. The large foot has become modified into a hatchet-shaped structure that is used in digging efficiently through the sediments. As relatively sedentary animals, pelecypods have lost their heads (no pun intended), and their enlarged gills serve as excellent filter-feeding structures. Although most clams have a burrowing mode of life, many live on the substrate,

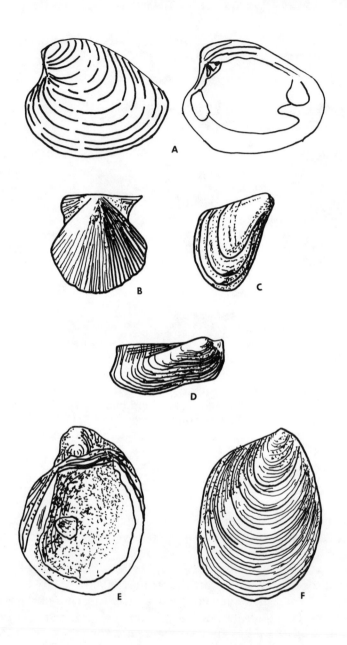

Pelecypoda (all but A from New Mexico). A, modern pelecypod, external and internal views, x 1; B, *Aviculopecten* (Pennsylvanian-Permian), x 1; C, *Myalina* (Pennsylvanian-Permian), x 1; D, *Parallelodon* (Pennsylvanian-Permian), x 1; E, *Pycnodonte*, an oyster, showing interior of shell (Cretaceous), x 1; F, *Inoceramus* (Cretaceous), x 1.

and some, like oysters, attach to it by cementing one of the shells to a hard object (such as the shell of another organism). A few pelecypods, like pectinids, are able to swim for short distances by clapping their valves together. The surface of pelecypod shells show concentric growth lines and many additional forms of ornamentation may be present. The shells of most pelecypods are roughly circular to elliptical in shape, but a few are triangular, some are very elongate, and one group of extinct Cretaceous pelecypods (the rudistids) enlarged one shell into a cone and reduced the other into a small cap, thus converging toward solitary corals in skeletal form.

Pelecypods are minor elements of Paleozoic marine faunas in New Mexico, but become abundant and dominant fossils in Cretaceous marine deposits. Most pelecypods are marine, but freshwater forms are moderately common in stream deposits in the Cretaceous and early Tertiary of northwestern New Mexico, and in the Plio-Pleistocene of eastern New Mexico.

Cephalopods. Cephalopods, represented today mainly by unshelled forms such as octopuses and squids, were more abundant in the past, when groups with prominent external shells like ammonoids and nautiloids were common. These fossil cephalopods had a single, multichambered shell that was planispirally coiled (coiled in one plane), a straight cone, or just about any shape between these two extremes. The soft body of the cephalopod was confined to the last chamber, and had the foot modified into numerous tentacles; the earlier chambers were filled with a gas-fluid mixture that gave the shell buoyancy. Each chamber is walled off from successive ones by shell material, and the intersection of these walls (septa) with the outside shell surface produced lines or sutures. These sutures are gently arcuate in nautiloids, but may be extremely complex in ammonoids. Unlike most other groups of molluscs, cephalopods adopted a predaceous swimming mode of life restricted to marine environments. Nautiloids are especially conspicuous in New Mexico Ordovician rocks, but become minor elements of marine assemblages in the middle Paleozoic; only one genus of nautiloid (the "chambered nautilus") has survived to the present. Ammonoids evolved from nautiloids in the Devonian, are uncommon in New Mexico Paleozoic strata, but become incredibly diverse and abundant in the Mesozoic. Cretaceous marine deposits in New Mexico have many dozens of different ammonoid species, but the group rather suddenly became extinct at the end of that period. Fossil ammonoids

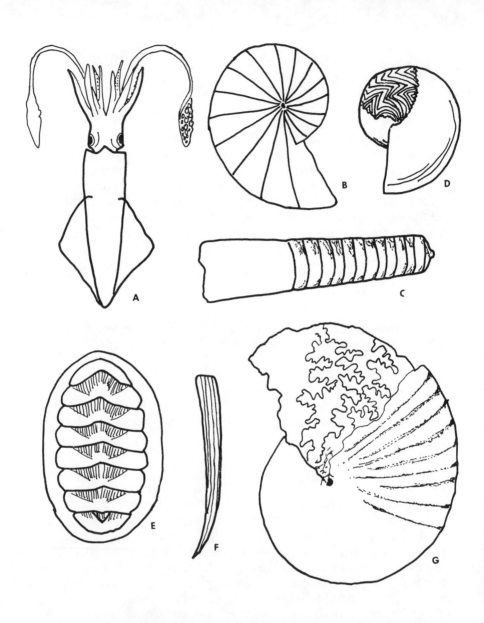

Cephalopoda and minor mollusc classes. A, modern squid, lacking an external shell, x 1; B, *Eutrephoceras*, a coiled nautiloid cephalopod, showing typical straight to gently curved sutures (Cretaceous), x .3; C, generalized uncoiled nautiloid (Ordovician), x 1; D, *Gonioloboceras*, an ammonoid with simple sutures shown on part of the shell (Pennsylvanian), x .5; E, generalized modern polyplacophoran, showing 8 overlapping plates, x 1; F, *Dentalium*, a scaphopod (Cretaceous), x 1; G, *Coilopoceras*, an advanced ammonoid with complex sutures, drawn in only near aperture (Cretaceous), x .5.

are frequently handsome and impressive fossils, with some New Mexico species reaching diameters of 75 cm or more.

Other Molluscs. Two other classes of molluscs are also occasionally found in New Mexico—Polyplacophorans (chitons), and Scaphopods ("tusk shells"). Polyplacophorans have a line of eight overlapping chevron-shaped shells; isolated chiton shells have been found in Ordovician, Pennsylvanian, and Permian rocks in New Mexico. Scaphopods have a single curved conical shell and are present in Pennsylvanian, Permian, and Cretaceous formations in the state, sometimes in surprisingly large numbers locally.

Arthropods

In terms of number of species, arthropods are the most successful group of animals; they constitute about three-quarters of all types of organisms living today and are found in almost every habitat or environment on earth. The phylum as a whole consists of segmented animals having jointed appendages and a flexible organic (chitinous) exoskeleton that is shed periodically as an individual grows. In some arthropods, such as crustaceans and trilobites, the chitinous exoskeleton is partially mineralized, making it stiffer and better able to become fossilized, but as a general rule arthropods are not preserved in the fossil record in anywhere near the abundance that they must have enjoyed when they were living. Arthropods are very advanced animals, with highly developed muscular, nervous, and sensory systems. A characteristic feature of arthropods are compound eyes, composed of numerous facets; these may be seen on especially well-preserved specimens. Some arthropods, such as bees and ants, also have complex social and behavioral interactions, but these are largely genetically programmed rather than learned. The main kinds of arthropods found in New Mexico rocks follow.

Trilobites. This extinct group is characterized by bodies composed of a head (cephalon), often with large dorsal compound eyes, a thorax of numerous narrow segments, and a tail (pygidium). Because of the large number of segments, trilobites are subject to disarticulation, and more fragments of trilobites (an isolated head or tail, for example) are found than complete specimens. Trilobites were small animals, usually less than 2–3 cm long, and most types spent their lives crawling on the sediments eating detritus.

THE MAIN FOSSIL GROUPS

Though they are not often preserved, each thoracic segment had a pair of walking legs underneath. Found only in marine environments, trilobites are major components of Cambrian and Ordovician communities, but they decline after this to become rather rare by the Pennsylvanian and extinct by the end of the Permian.

Eurypterids. Also extinct, eurypterids ("water scorpions") are rare fossils; nearly all reported New Mexico specimens come from a single Pennsylvanian-Permian locality west of Los Lunas. Eurypterids are remotely related to arachnids, which include modern land scorpions, and the general form of the body does indeed resemble that of a large scorpion. The eurypterid body consists of a large rounded head, twelve body segments, a long spiny tail, and six pairs of appendages attached to the underside of the head. The last pair is modified into elongate paddles, which allowed them to swim slowly through the water; the other appendages were walking legs or food manipulation structures. Though starting out as marine creatures, eurypterids became restricted to fresh- or brackish-water environments, and the late Paleozoic New Mexico forms are found in swamp deposits along with large numbers of plant leaves and insects.

Arachnids. This group includes the spiders and scorpions. They are very numerous and successful today, and their fossil record goes back to the middle Paleozoic, but so far no fossils have been reported from New Mexico.

Crustaceans. Possibly related to trilobites, crustaceans are rare in New Mexico's fossil record, but are the most diverse and successful group of water-dwelling arthropods alive today. Shrimp, lobsters, and crabs are perhaps the most familiar modern crustaceans, and fossils of these groups have occasionally been found in Pennsylvanian and Cretaceous marine sediments within the state. The knobby bifurcating burrows of a shallow-marine shrimp are quite common in some of the Upper Cretaceous shoreline sandstones in northern New Mexico. More common as fossils are the ostracods, a group of very small crustaceans characterized by two articulated calcareous shells that cause them to superficially resemble miniature clams. Most ostracods are less than 2 mm in length, and thus are really microfossils, but they are locally abundant (though hard to see) in marine and fresh-water sediments from the Ordovician on in New Mexico.

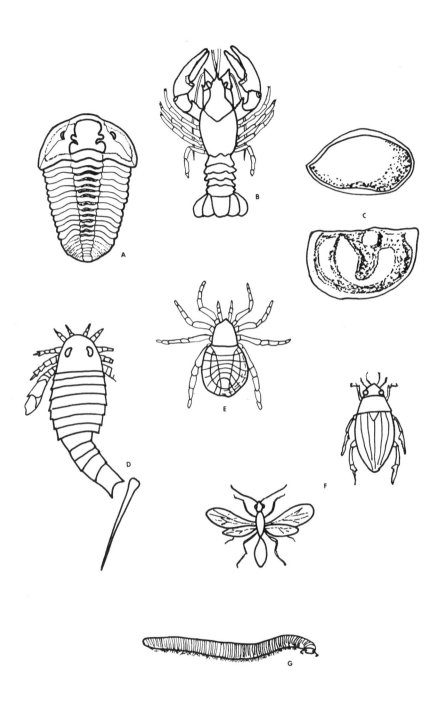

Arthropoda. A, Trilobite, x .75; B, Crustacean (lobster), x .3; C, Crustacean (Ostracods), x 15; D, Eurypterid, x .375; E, Arachnid (spider), x 2; F, Insects, x 1.5; G, Millipede, x 1.5.

THE MAIN FOSSIL GROUPS

Insects. Insects are the most abundant of modern arthropods, and are the only invertebrates that have developed the ability to fly. Presumably they were equally abundant in the past, but the fossil record of insects is extremely poor. New Mexico's insect record is limited to a few Pennsylvanian localities, particularly the swamp sediments that yield the eurypterids mentioned above, and to some supposed insect borings in Paleocene petrified wood. Many of the Pennsylvanian insects are cockroaches, a group, unfortunately, still very common today, but the eurypterid locality near Los Lunas has also produced some real giants—dragonflylike forms with a wingspan of 30 cm or more. Related to insects are the millipedes and centipedes, highly segmented terrestrial arthropods with many legs; a few millipedes have also come from the Pennsylvanian eurypterid locality.

Echinoderms

Echinoderms, or "spiny-skinned animals," include seastars, sea urchins, crinoids, and other pentamerously symmetrical invertebrates. The calcareous skeleton of echinoderms consists of hundreds of articulated plates or ossicles that tend to become disarticulated after death, causing most fossil enchinoderms to be represented by incomplete remains. The modern echinoderm groups all appear in the Ordovician; there are in addition numerous groups that flourished during the Paleozoic, and became extinct before the end of that era. The most abundant fossil echinoderms are the crinoids, sometimes called "sea lilies" because of their superficial resemblance to plants. Crinoids consist of a budlike calyx composed of many polygonal plates and surmounted by numerous, often highly branched, arms utilized in filter-feeding. The calyx is situated at the end of a long stem, composed of hundreds of small poker chiplike segments, that was attached to the sea floor. Calyces are rarely found as fossils; an exception occurs in the Mississippian beds near the little town of Lake Valley, New Mexico, which have yielded large numbers of complete and exquisitely preserved crinoid calyces. Parts of crinoid stems, on the other hand, are among the most abundant fossils found in Paleozoic marine sediments. Blastoids, an extinct group of crinoidlike echinoderms, have a calyx consisting of only thirteen plates, lack arms, and have five prominent petal-shaped ambulacral areas projecting from the top down along the sides of the calyx. Blastoids were most abundant during the Mississippian Period, but are uncommon in New Mexico rocks of that age.

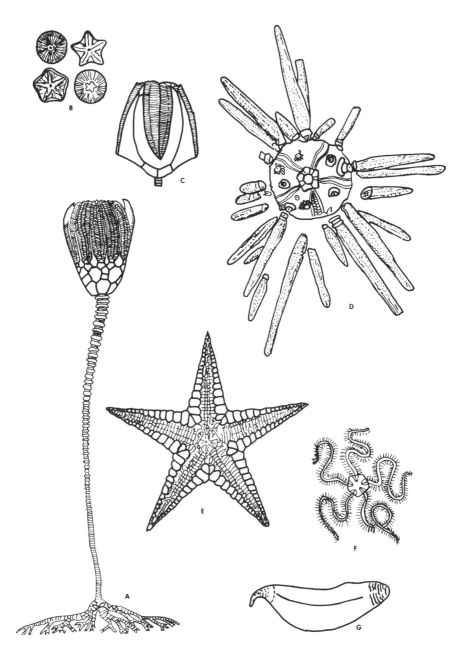

Echinodermata. A, Crinoid, x .75; B, Several different types of crinoid stem segments, x .75–1.5; C, Blastoid calyx, x 1; D, Echinoid (sea urchin), showing spherical test and some of the spines, x .75; E, Asteroid (seastar), x .75; F, Ophiuroid (brittle star), x .75; G, Holothurian ("sea cucumber"), x .75.

THE MAIN FOSSIL GROUPS

Graptolites. A, Dendroid graptolite (Ordovician), x .75; B, Two-branched graptoloid graptolite (Ordovician) x 1.5; C, D, Single-branched graptoloid graptolites (Ordovician-Silurian), x 1.5.

Echinoderms capable of locomotion include seastars and echinoids. Seastars are common today along marine shorelines, but are very rare as fossils—the only reported occurrences in New Mexico being in a Mississippian bed in the San Andres Mountains and imprints in a Cretaceous sandstone from the Rio Puerco Valley. Echinoids, or sea urchins, generally have a spherical shell with numerous long sharp spines projecting from the surface. Complete shells are very rare in Paleozoic rocks, but isolated spines are not uncommon; in the Cretaceous of southern New Mexico, however, well-preserved complete shells are fairly abundant in some strata. Advanced "irregular" echinoids tend to have flattened shells and bilateral rather than pentamerous symmetry; sand dollars are the most extreme example of these trends. Both seastars and echinoids move slowly over the substrate, with seastars feeding mainly on pelecypods and echinoids rasping algae and other organisms from rocks by means of strong plates situated in the mouth.

Several other echinoderm groups are worth mentioning, though their remains are seldom found in New Mexico. Ophiuroids, or "brittle stars," resemble seastars, but have long snaky arms; only isolated ossicles have been reported from New Mexico Paleozoic rocks. Holothurians, or "sea cucumbers," are abundant in modern oceans, but are seldom found as fossils because their calcareous skeleton has become reduced to microscopic hook- or disc-shaped spicules isolated in the thick skin of the animal, giving holothurians the appearance of fat soft worms. Cystoids, an extinct stalked group similar to crinoids, are known in New Mexico from isolated polygonal pore-bearing calyx plates in lower and middle Paleozoic deposits.

INVERTEBRATE FOSSILS

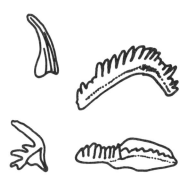

Conodonts. Several genera, showing simple forms with one or a few cusps to left (Ordovician-Silurian), and more advanced forms with numerous cusps and flattened platforms to right (Mississippian). All are magnified about 20 times.

Graptolites

Graptolites are extinct members of a minor phylum called the Hemichordata. They were colonial animals with a small delicate organic skeleton, and are usually found preserved as carbonaceous impressions on the bedding planes of dark gray or black marine shales. Their colonies may either be highly branching or consist of four, two, or one branches having saw-toothed margins on one or both sides of the branch. Graptolites are considered to have been planktonic organisms and have a geologic range from the Cambrian to Mississippian. Few graptolites have been found in New Mexico, but highly branched forms are known from Early Ordovician rocks in southern New Mexico.

Conodonts

These nearly microscopic skeletal elements resemble miniature teeth in having one to many sharp cusps projecting from an elongate or flattened platformlike base. Conodonts are a paleontological enigma because, although they are relatively common in many Paleozoic and early Mesozoic marine sediments, the nature of the animal from which they came is still a mystery. Conodonts are composed of calcium phosphate, the same material that makes up vertebrate bones and teeth, but these fossils are constructed in such a way that they could not have been teeth. No wear facets are ever present and there are signs that some broken conodonts have been repaired, something that true teeth never show. Until a good fossil of the "conodont animal" is discovered, conodont affinities will remain uncertain. The best guess, however, is that they were skeletal parts of a mainly soft-bodied swimming or floating organism, probably some kind of primitive extinct chordate. Conodonts are present, though nearly impossible to see with the naked eye because of their small size, in rocks of Ordovician through Permian age in New Mexico.

Vertebrate Fossils

Vertebrates form the main subdivision of the phylum Chordata, and include those organisms possessing endoskeletons composed of the mineral hydroxyapatite (basically calcium phosphate) or its organic precursor, cartilage, in the form of numerous articulated bones. Complete fossil skeletons of vertebrates are uncommon because the bones readily become disarticulated during transportation and burial before fossilization can occur. The group as a whole is very advanced in the complexity of its soft parts, especially the nervous and sensory systems, and most vertebrates are active animals, with good locomotion capabilities and large brains. Vertebrates include the largest animals that have ever lived, in water as well as on land. There are nine classes of vertebrates; these are summarized below, with illustrations of typical representatives of each class.

Agnathans, Acanthodians, and Placoderms

These three classes include armored primitive fishlike vertebrates, with a poor to absent fossil record in New Mexico. Agnathans, the only one of these classes that is not extinct (modern lampreys and hagfish are agnathans), are the most primitive and geologically earliest vertebrates, appearing in the Late Cambrian, and are the only vertebrates to lack true jaws and (as a primary characteristic) paired lateral appendages. Acanthodians and placoderms were generally predaceous, had variable numbers of long spines, and some reached large sizes. All of these classes reached their evolutionary apex in the Devonian, and dwindled rapidly afterward. Acanthodian spines are known from the Devonian of New Mexico, and fragmentary skeletons have been found in Pennsylvanian rocks of the Manzano Mountains.

Chondrichthyans

This class includes the sharks and rays, vertebrates with a cartilaginous endoskeleton, several gill slits on both sides of the head, an elongate upper lobe in the tail fin, and scales that closely resemble teeth, both in appearance and genesis. With a fossil record that goes back to the Devonian, this group is successful today; many sharks are fast-swimming predaceous carnivores and most rays are invertebrate-eating bottom dwellers. Because of the lack of skeletal bone, chondrichthyans are rather poorly known as fossils, except for their teeth and scales. Essentially modern sharks arose and flourished in the Cretaceous, and fossil shark and ray teeth are locally common in many New Mexico marine sandstones of that age. The teeth of more primitive sharks are uncommon, but occasionally found in rocks of Devonian to Permian age around the state.

Osteichthyans

Osteichthyans, or "bony fish," include most of the familiar fish in freshwater and marine environments today, and constitute the most successful vertebrate group, in terms of number of extant species. Bony fish are characterized by a single gill slit on each side of the head, and by the presence of flat rhomboidal to circular overlapping scales set into the skin. Primitive members had very thick armorlike scales and an elongate upper tail fin lobe, but modern forms generally have very thin, sometimes transparent bony scales and a symmetrical tail. Bony fish have adapted to virtually all aqueous habitats (a few are able to live on land for extended periods) and through their history have developed a tremendous variety of body forms, from snakelike (eels) to almost totally round (ocean sunfish).

The fossil record of bony fish began in the Devonian (they probably evolved from acanthodians), and highly advanced forms rose quickly to dominance in the Cretaceous, a position they still hold today. Most bony fish are "ray-finned," their fins being supported by numerous needlelike bones, but some possessed lobelike lateral fins and functioning lungs, and gave rise in the Late Devonian to terrestrial vertebrates (amphibians). Lobe-finned fish are now limited to a few species of lungfish and coelacanths. Fossils of bony fish are not common in New Mexico rocks—teeth and scales are the most frequently found remains—but complete skeletons have been found in the Pennsylvanian and Jurassic rocks of the central part of the state.

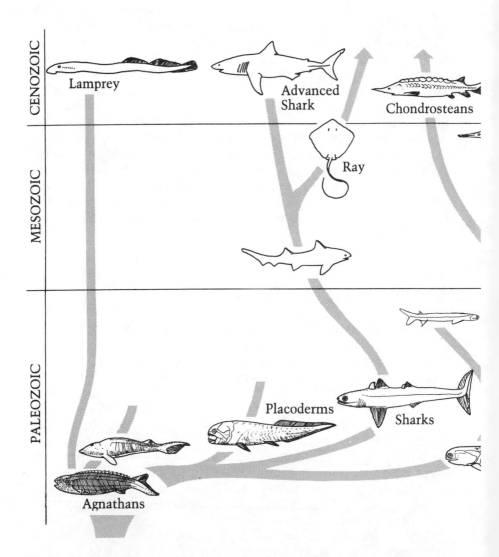

Relationships of major groups of aquatic vertebrates. Diagrams of fishes are not necessarily to the same scale, and only a few of the scales on each fish have been drawn in. Advanced sharks, rays, and teleost fish are by far the most abundant groups today; lampreys, chondrosteans, holosteans, and coelacanths are each represented by

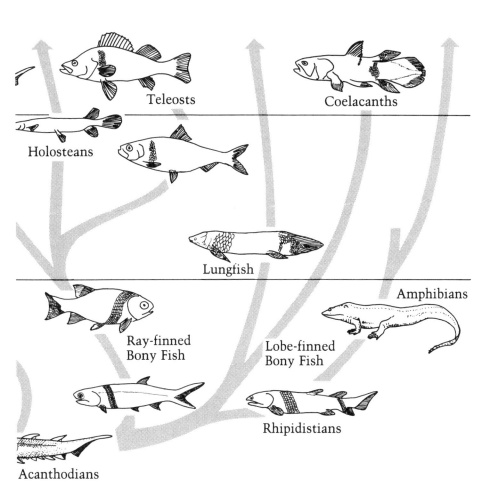

one or a few species. Amphibians evolved from rhipidistian lobe-finned fish in the Late Devonian; modern amphibians, such as frogs, toads, and salamanders, arose in the middle part of the Mesozoic Era.

THE MAIN FOSSIL GROUPS

Amphibians

These partially terrestrial vertebrates are represented today by frogs, toads, and salamanders. Amphibians lack scales, have legs rather than fins as adults, and are able to move around on land, but all amphibians are tied to bodies of water by the necessity of breeding and laying eggs in water (or at least in very moist conditions). Young amphibians (like tadpoles) function as fish, with gills and fins, before metamorphosing into adults. Amphibians developed from lobe-finned fish in the Late Devonian, and were the dominant terrestrial vertebrates until the Permian. Some late Paleozoic and early Mesozoic amphibians reached lengths of 4–5 meters and resembled stubby alligators. Excellent amphibian fossils have come from New Mexico rocks of Pennsylvanian, Permian, and Triassic age, the best deposit being an "amphibian graveyard" of numerous skulls and skeletons from the Triassic redbeds near Lamy. Fossil examples of the modern amphibian groups are very rare, but a few salamander bones have come from Late Cretaceous rocks in northwestern New Mexico.

Reptiles

Developing from amphibians in the Pennsylvanian Period, reptiles were freed from the necessity of remaining near bodies of water by developing an egg that contained its own aqueous environment, a large food supply, and a resistant outer covering. Colonization of a much broader range of terrestrial habitats than were open to amphibians made reptiles the dominant terrestrial vertebrates by the Permian, a position they were to hold for the next 200 million years. During this time reptiles diversified greatly, producing many groups now extinct in addition to the crocodilians, lizards, snakes, turtles, and tuataras that represent the modern remnants of the group. One early and successful line culminated in the Triassic with the development of mammals, while another led to the large and varied group called the archosaurs. Archosaurs, the major reptiles of the Mesozoic Era, include crocodilians, flying reptiles (pterosaurs), and the groups that have come to be known as dinosaurs.

Beginning with small, agile, bipedal, carnivorous forms like *Coelophysis* (New Mexico's state fossil), of which many skeletons are known from the Triassic redbeds near Ghost Ranch, New Mexico, dinosaurs developed in three main directions during the Jurassic and Cretaceous: (1) Bipedal carnivores of various sizes,

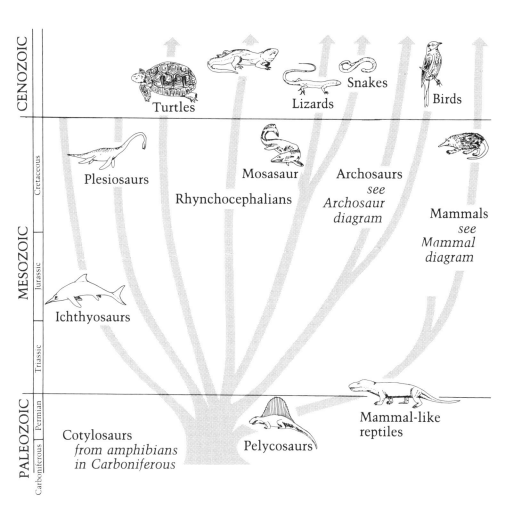

Relationships of major groups of reptiles and reptile derivatives (birds and mammals). Diagrams are not to the same scale; the various archosaur groups are portrayed in a separate diagram. Reptiles diversified considerably in the Late Paleozoic, and by the Triassic numerous very distinctive reptilian groups had developed, the most important of which are shown in this diagram. Mammals, a separate vertebrate class, evolved from mammallike reptiles in the Late Triassic. Birds, also a separate class, evolved from archosaurs in the Jurassic. Archosaurs, the most successful group of reptiles, dominated terrestrial environments through the Mesozoic but most archosaurs, including the dinosaurs, became extinct at the end of the Cretaceous.

THE MAIN FOSSIL GROUPS

from that of a turkey to gigantic forms like *Tyrannosaurus* and related genera that stood 6 meters high and measured up to 15 meters in length, and had saw-edged teeth up to 15 cm long; (2) herbivorous sauropods (*Camarasaurus* is an example), the largest land animals that have ever existed (some were 25 meters or more in length), characterized by absurdly small heads, extremely long necks and tails, and massive bodies held up by four columnar legs; and (3) "bird hipped" dinosaurs, a heterogeneous plant-eating group that includes the bipedal "duck-billed dinosaurs," large quadrupedal horned dinosaurs (ceratopsians), stegosaurs, with plates or spines on their backs, and the squat heavily armored ankylosaurs. Although they have been popularly portrayed as stupid, unsuccessful animals, dinosaurs were highly varied, they dominated terrestrial communities for nearly 150 million years, and they included types with a brain capacity larger than any modern vertebrates except mammals. Very probably some dinosaurs also developed the physiological mechanisms that allowed maintenance of an elevated body temperature; they became warm-blooded, as only mammals and birds are today.

Many other reptiles were successful during the Mesozoic. The pterosaurs, with broad membranous wings supported by the arms and little fingers of the hands, were the first vertebrates to develop the ability to fly, and they ranged in size from that of a crow to forms with a wingspan of about 17 meters—the largest flying creatures of all time. Several reptilian groups became fully adapted to an aquatic life: the streamlined dolphin-like ichthyosaurs, the ungainly long-necked plesiosaurs, and giant marine lizards (mosasaurs) among them. All of these groups of large reptiles became extinct at the end of the Cretaceous Period, but lizards and snakes, which were relatively minor elements of late Mesozoic faunas, survived and expanded to become the most numerous remaining reptiles today. Turtles, which first evolved in the Triassic, were successful in the Mesozoic and are still common in both terrestrial and aqueous environments.

Reptiles are among the most common vertebrate fossils in New Mexico rocks. Primitive reptiles, some of them closely related to the line that would ultimately lead to mammals, are found in Permian redbeds, and several different types of Triassic reptiles are also well represented in the north-central part of the state, including *Coelophysis*, one of the first dinosaurs, and phytosaurs, large amphibious crocodilelike predators. In Jurassic rocks, bone fragments and occasionally complete bones and partial skeletons come from the Morrison Formation, the unit that has magnificent dinosaur deposits in Colorado and Utah. The best deposits of fossil reptiles are in the northwestern part of New Mexico south of Farmington, where extensive badlands of Late Cretaceous age are exposed. Numerous types of dinosaurs, as well as abundant contemporaneous crocodilians, turtles, and rare lizards and snakes are constantly eroding out of these strata. The overlying Paleocene and Eocene beds, though lacking dinosaurs, which became extinct at the end of the Cretaceous, yield crocodilian, turtle, and lizard bones. A few reptile fossils, mainly turtles and tortoises, are found in Miocene-Pliocene and Pleistocene rocks in the northern part of New Mexico.

Birds

Birds developed from archosaurian reptiles in the Jurassic Period, and the first bird, were it not for the presence of feathers, would be classified as a small dinosaur. Birds, of course, are one of the few groups of animals to have developed the ability to fly, and many of their distinctive features are adaptations related to their flying mode of life. Along with feathers, which are elaborate developments of reptilian scales, birds have their front appendages modified into wings, are warm-blooded (flying requires the constant generation of much metabolic energy), and have hollow bones. In addition, bird jaws are toothless, and modified into beaks, and birds lay eggs, just as their reptilian forebears did. A very successful group today, birds have left a poorer fossil record than most other vertebrate classes, primarily because of the fragility and small size of their skeletons and because most birds live on land, where depositional conditions conducive to burial and fossilization are not common. In New Mexico, fossil birds are rare. A fragment of a wingbone was recently discovered in marine Cretaceous strata, some skeletal elements of a gigantic flightless bird are known from the Eocene, and a vulturelike form from the Mio-Pliocene badlands are the extent of the record in northern New Mexico. Several types of Pleistocene to Recent fossil birds, including an extinct roadrunner and a giant condor, have been found in several caves in southern New Mexico.

Mammals

Mammals evolved from "mammallike" reptiles during the Triassic by gradually developing a set of new structural and physiological features; the transition was so subtle that there are

Relationships of major groups of archosaurs. Diagrams not to same scale. Saurischian ("lizard-hipped") dinosaurs include the sauropods, carnosaurs, and coelurosaurs. Ornithischian ("bird-hipped") dinosaurs include the ceratopsians, ornithopods (of which the hadrosaurs are

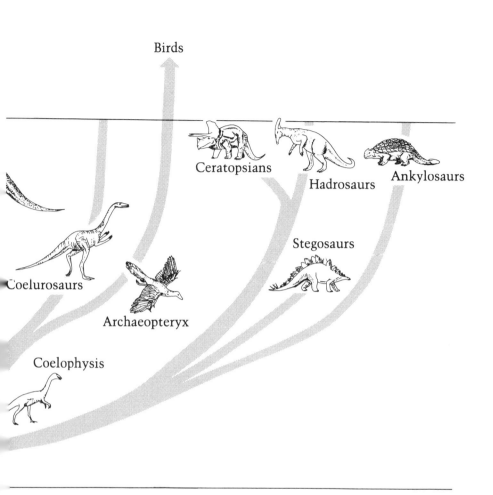

the most important group), stegosaurs, and ankylosaurs. Birds are included among the archosaurs by some paleontologists; traditionally they have been considered a separate class of vertebrates.

THE MAIN FOSSIL GROUPS

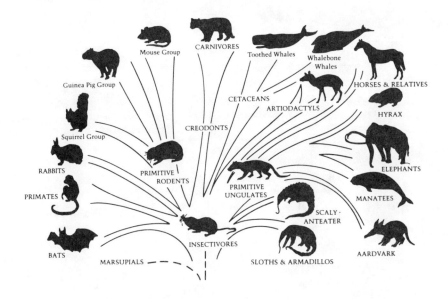

Relationships of major mammal groups.

some fossils that represent perfect intermediates between reptiles and mammals. The presence of teeth differentiated into several types (for example, canines, incisors molars) having different functions, in contrast to the usual undifferentiated conical teeth of reptiles and other lower vertebrates, is one of the major advances shown by mammals. Mammals are also characterized by having only one bone in the lower jaw (reptiles have several), and three small ear bones (as opposed to one in reptiles). More subtle advances, however, such as the physiological maintenance of an elevated body temperature, the presence of hair or fur, the bearing of live young instead of laying eggs, and the subsequent suckling of the young on specialized nutrient-providing glands coupled with increased parental care are perhaps more important characteristics of the class Mammalia. Mammals also have the largest brains and highest intelligence of any animals.

Though mammals developed at about the same time as the archosaurian reptiles, the reptiles prevailed through the Mesozoic, while primitive mammals remained very small (and probably spent a lot of their time staying away from the large variety of reptiles). The earliest mammal fossils in New Mexico are Late Cretaceous forms from the dinosaur beds of northwestern New Mexico, represented by minute teeth and jaws that indicate a body size no larger than that of a small cat. After the dinosaurs became extinct, however, mammals quickly expanded and diversified to become the dominant terrestrial vertebrates, and the Paleocene and Eocene exposures in northwestern New Mexico have some of the richest deposits of primitive mammals known in the world. Most of these primitive mammals were small (though one group reached the size of a cow), and the majority belong to extinct orders that represent early and ultimately unsuccessful evolutionary experiments. A few of the early mammals, however, mark the beginnings of modern mammalian groups; for example, the first horse, a little animal about the size of a fox, is relatively common in the 50-million-year-old Eocene deposits of northwestern New Mexico.

Through the Cenozoic mammals evolved rapidly, and by Miocene-Pliocene time, about 5–15 million years ago, many essentially modern mammals had developed, such as cats, dogs, camels, elephants, horses, antelopes, rhinoceroses, rabbits, and rodents. The badlands around Santa Fe and Espanola have produced thousands of mammal fossils of this period, and very recent rocks of Pleistocene age in many parts of the state contain numerous ice-age mammals, many still extant but some recently extinct. At a few sites, such as Folsom and Clovis, the remains of extinct mammals are found in association with the artifacts of Paleo-Indians. It was at this time, about 15,000 years ago, that the most advanced mammals, humans, entered New Mexico.

Plants

The fossil record of plants begins in the Precambrian, long before the existence of animals. Most Precambrian fossils are primitive unicellular photosynthesizing organisms assigned to the Kingdom Monera (bacteria and blue-green algae); true plants probably developed about 1.5 billion years ago. A major effect of monerans and plants on the early earth was gradually to increase the amount of oxygen available in the atmosphere and in bodies of water on the earth's surface, allowing the eventual development

THE MAIN FOSSIL GROUPS

of oxygen-using animals toward the end of the Precambrian. The earth's present atmosphere is unique among all known planetary atmospheres in its high (about 20 percent) oxygen content, due largely to the activities of photosynthesizing organisms during the past three billion years. In addition, from the Precambrian to the present, animals have depended on plants directly or indirectly for food. Because only plants have the ability to transform the sun's energy into organic material, plants are at the base of both aqueous and terrestrial food chains.

Most plants lack mineralized skeletal parts, and for this reason are underrepresented in the fossil record. Generally, plant fossils, especially leaves, are preserved as carbonaceous remains or impressions in fine-grained sedimentary rocks; petrifaction of stems and trunks is another common way in which plants have been fossilized. In situations where burial has been gentle and the sediments have not been significantly disturbed during lithification, plant fossils may be exquisitely preserved, showing extremely fine details of leaf venation and other structures. During the last 350 million years plants have been abundant constituents of most terrestrial environments, and it is here that the most advanced plants so familiar to us all have thrived. Plants are also abundant in the oceans, but these are primarily relatively primitive forms (algae) that are limited to about the upper 100–200 meters because of their need for sunlight, which diminishes drastically below these depths.

There are about fifteen major plant groups, equivalent to animal phyla, but we will here consider three general plant types—algae, lower vascular plants, and higher vascular plants. Algae include many plants, especially several different types of seaweed, that have simple reproductive structures and lack true leaves, roots, and internal vascular structures. Nearly all algae live in water, and they are the only plants known until the Silurian Period. Fossils of most algae are very uncommon, but some types have secreted calcareous skeletons since the Cambrian and their fossils are relatively well known. Calcareous algae have always been significant constituents of reefs and in many cases they predominate in these structures, being more important than the more familiar reef-building organisms such as corals. Calcareous algae have a wide variety of forms, but many are preserved as undulating laminae, crinkly "potato-chip"-like structures, and as fragments from originally branched and jointed skeletons. Algae with no skeletons have been important in trapping and binding sediment, sometimes producing small mounds that are preserved in the rock record. Because algae are ubiquitous in modern oceans, we assume that they were equally so in the past, even though evidence of their presence is absent in most marine fossil deposits.

In the Silurian, the first land plants appeared, and the inherent problems of living out of the water necessitated the development of a series of new structures. Water-dwelling plants can absorb nutrients and water through all parts of their surface area, but land plants require roots to absorb these materials from the soil, where they are concentrated. Specialized internal vascular structures are also required to convey water and nutrients from the roots to all parts of the plant. Sturdy stems and branches are needed for support on land, whereas the water itself supports aqueous plants. Changes in reproductive structures developed to compensate for the lack of a surrounding aqueous medium that facilitates fertilization. Resistant cuticles evolved in land plants to protect against drying out, a problem that water-dwelling plants do not have.

These adaptations developed gradually in early land plants, and these plants initially were probably restricted to moist environments along the banks of streams and lakes or marine shorelines. The earliest vascular plants lacked leaves and true roots, but internal vascular structures were present; they consisted of spore-bearing upright shoots, sometimes covered with leaflike scales, that rose from a ground-hugging rhizome. They are rare as fossils. From them, however, during the Devonian several more advanced groups evolved, producing the first forests by the end of that period, and increasing in abundance and diversity to culminate in the great, dense, continent-spanning forests of the Mississippian through Permian periods. Land plants were so abundant during this time that their fossilized remains account for the majority of coal mined in the world today, and led European geologists to name the time of their greatest success as the Carboniferous Period.

Carboniferous forests were dominated by plant groups that have few representatives today, and these forests would seem alien to a hypothetical modern visitor. Lycopods ("club mosses") such as *Lepidodendron* reached heights of 30 meters or more, had scale-like leaves on the branches and trunks, and possessed distinctive diamond-shaped patterns on the long trunks that call to mind automobile tire tracks. Sphenophytes ("horsetails" or "scouring rushes") like *Calamites* had jointed cylindrical central trunks with a fringe of simple leaves or leafy branches at intervals around them; some were more vinelike and possessed circlets of leaves

THE MAIN FOSSIL GROUPS

A Late Silurian scene, showing very primitive terrestrial vegetation, concentrated along bodies of fresh water. *(From a painting by Z. Burian)*

along a thin flexible stem. Ferns, a group that is still moderately common in moist environments today, were represented by hundreds of different types, some reaching the proportions of trees, but all characterized by distinctive elliptical leaves with complex and highly varied venation patterns. All of these lower vascular plants reproduced by spores and lacked the characteristic feature of advanced vascular plants—seeds.

Seeds are somewhat analogous to eggs in animals, providing a food supply and tough outer covering for the early sustenance and protection of young plants. Seeds probably first appeared during the Carboniferous on plants with fernlike foliage ("seed ferns"), but they reached their highest development in the two most advanced groups of plants, the gymnosperms and angiosperms. These two groups dominate terrestrial vegetation today and include all of the plants used by mankind for such things as food,

PLANTS

A Pennsylvanian coal forest composed of lush vegetation that included several types of large trees. Left foreground, *Lepidodendron*; left center, *Calamites*; right foreground, tree fern; far right, *Cordaites*, an early gymnosperm. Although conditions in New Mexico were not suitable for the formation of thick coal beds during the Pennsylvanian, all of the plants pointed out above have been found in the state. Dragonflylike insects with wingspans of up to 50 cm lived in the Pennsylvanian; a few have been found in New Mexico. *(From a painting by Z. Burian)*

lumber, and landscaping. The gymnosperms have their seeds "naked" in cones, and the group includes the familiar conifers (pine, spruce, juniper, and cypress trees) as well as less common plants like cycads, ginkgos, and *Ephedra* (the "Mormon tea" of Southwestern deserts). Early gymnosperms were common in the late Paleozoic; many have straplike leaves (*Cordaites*), but others show the needlelike leaves typical of conifers. Gymnosperms are the most important Mesozoic plants, and some concentrations of gymnosperm fossils, such as those at Petrified Forest National Park in northeastern Arizona, are petrified in bright hues of red, yellow, and orange.

Angiosperms, or flowering plants, arose in the Cretaceous Period, and today are by far the most abundant and varied of terrestrial plants. They are characterized by complex reproductive structures that include flowers and fruit (a fleshy covering around the seeds), and often by wide, highly veined leaves of many shapes. Angiosperms have diversified into a wide variety of different forms, ranging from grasses to palm and fruit trees to water lilies, and a few types, such as "turtle grass" and mangrove trees, are even able to live in fully marine conditions, something no other terrestrial plants are able to do. The rapid evolution of angiosperms in the Late Cretaceous resulted in a significant change in the composition of many forests, from primarily gymnosperm to primarily angiosperm; this and the evolution and expansion of grasses in the early Tertiary strongly influenced the evolution of many herbivorous mammal groups. Grasses contain silica, which can rapidly wear down the teeth of animals that use them for food; mammals that eat grass developed much more resistant teeth than those that subsist on softer leafy vegetation. In addition, the replacement of forests by extensive grasslands has had important implications in the evolution of locomotion and protection structures in some mammals.

The record of plant fossils in New Mexico is sporadic. Lower vascular plants are known from some Mississippian outcrops in southwestern New Mexico; they are occasionally present in marine deposits and represent terrestrial plants that were washed out to sea. Remnants of Pennsylvanian coal forests are moderately common in many parts of the state; wherever thick deposits of Pennsylvanian strata that include gray and black shales are present, some plant fossils are to be expected, though often they are in a highly fragmented state. Well-preserved Pennsylvanian ferns and other plants have come from the Manzano, Nacimiento, and Sangre de Cristo mountains, Lucero Mesa, and near Socorro, among other localities. Early Permian red sandstones like the Abo Formation occasionally have plant fossils, too, but they are generally not as common or as well preserved as in Pennsylvanian exposures. Triassic red sandstones have yielded mainly cycads and ferns near Fort Wingate, Abiquiu, Santa Rosa, and Tucumcari; the brightly colored petrified trunks of Chinle Formation conifers just across the border in northeastern Arizona's Petrified Forest have already been mentioned.

Terrestrial Jurassic, Cretaceous, Paleocene, and Eocene deposits have locally abundant concentrations of petrified wood throughout their outcrop areas. Many excellent (mainly angiosperm) leaf localities are known in the Late Cretaceous and Paleocene coal beds south of Farmington, around Cuba, and all over the Vermejo Park to Raton area. A significant number of species from these localities is apparently unique to New Mexico. More recent rocks

Lower vascular plants (all found in New Mexico). A, Reconstruction of *Lepidodendron*, a large lycopod up to 30 meters high; B, Part of trunk of *Sigillaria*, a lycopod, x .24; C, Detail of trunk of *Lepidodendron*, showing diamond-shaped leaf scars, x .25; D, Reconstruction of *Calamites*, a large (10-meter-high) sphenophyte; E, A tree fern, original about 15 meters high; F, *Neuropteris*, a seed fern, x .3; G, *Annularia*, branch foliage of *Calamites*, x .6; H, Cast of *Calamites* trunk segment, x .3; I, *Dryopteris*, a fern, x .3; J, *Mariopteris*, a seed fern, x .3; K, *Pecopteris*, a fern, x .3. All plants shown are of Pennsylvanian age except I, which is Paleocene.

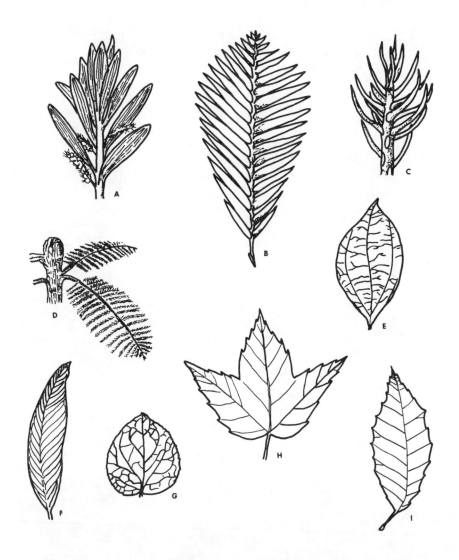

Gymnosperms and angiosperms. A, *Cordaites*, an early gymnosperm (Pennsylvanian), x .2; B, *Zamites*, a cycad (Triassic), x .25; C, *Araucaria*, a conifer (Norfolk Island Pine) (Cretaceous–early Tertiary), x 1.5; D, *Walchia*, an early conifer (Pennsylvanian-Permian), x .2; E, *Cinnamomum*, an angiosperm (Cretaceous-Recent), x .25; F, *Magnolia*, an angiosperm (Late Cretaceous–Recent), x .2; G, *Cercidiphyllum*, katsura leaf, an angiosperm (Late Cretaceous–Recent), x .25; H, *Acer*, maple leaf, an angiosperm (early Tertiary–Recent), x .4; I, *Quercus*, oak leaf, an angiosperm (Late Cretaceous–Recent), x .25. Though many Late Cretaceous to early Tertiary angiosperm leaves are classified within modern genera, the resemblances in some cases are only superficial and no close relationships exist.

(Miocene to Pleistocene age) have not produced many plant fossils; perhaps the most unusual forms are Pleistocene plants from the Rincon Hills in southern New Mexico that have had their roots and tubers transformed into black and white colored opal. Microscopic plant fossils, such as spores and pollen, are known to be abundant in some beds in rocks as old as the Pennsylvanian in New Mexico, and several dozen species have been reported from the Late Cretaceous–Paleocene sequence in the San Juan Basin of northwestern New Mexico. Because they require special techniques to remove them from the rocks, and are too small to see with the naked eye, fossil spores and pollen generally are not observed except by paleontologists who are specifically looking for them.

3

THE NEW MEXICO FOSSIL RECORD

Chapter 3 is a guide to New Mexico fossils of each geologic period represented by rocks in the state. In the discussion for each period is a summary of important paleontologic events during that time, a map showing the distribution of outcrops, and a general description of the types of fossils one is likely to find in strata in New Mexico. Illustrations of common and characteristic genera are provided to facilitate identification of fossils collected or observed.

In using this identification guide, one should remember that about 6,000 different kinds of fossils have been reported from New Mexico, and, therefore, the illustrations include only a fraction of the total number of different genera that are known to be present in the state. To identify fossils that are not portrayed in this book, it is necessary to go to the technical literature or to a paleontologist, who will be able to help with identification. Because there are variations in the structure and form of species included within a given genus (the most obvious being the increase in size of individuals with age), fossils belonging to the genus may not look exactly like the illustration for it.

Not all the specimens shown on a page are necessarily drawn to the same scale. Most figures of invertebrates are roughly life-sized, but in some cases the illustrations of large fossils have had to be reduced for efficient use of space, and those of small fossils enlarged to show detail. The degree to which a particular illustration departs from the actual size of a typical specimen is indicated in the captions; for example, ×2 means that the illustration is twice the actual size of the fossil.

Figures of complete specimens are provided for most invertebrates because these fossils are often found that way. Vertebrate fossils, on the other hand, are rarely complete, and so, although some common diagnostic skeletal elements of vertebrates are illustrated in the appropriate places, these have been supplemented with reconstructions of entire animals, to aid the reader in visualizing the appearance of common vertebrates that populated the state at various times. Plants also are rarely found in a complete state; most of the figures of plants are for commonly found parts, such as leaves.

Distribution of Fossiliferous Rocks in New Mexico

Sedimentary rocks containing fossils are exposed over most of New Mexico's area, but the distribution of rocks of various ages is not constant from region to region within the state. Thus, for example, rocks of Cambrian through Devonian age are limited to the southern third of New Mexico, whereas nearly all early Tertiary (Paleocene and Eocene) sediments and fossils are restricted to the northern part of the state. In addition, there is a great variation in the thicknesses of sedimentary rocks representing the geologic periods. The total thicknesses of Cambrian or Jurassic rocks, for example, are only a small fraction of the total thicknesses of Pennsylvanian or Cretaceous rocks within the state. These variations are a result of the multitudinous events that constitute New Mexico's geologic history. They are mainly influenced by the nature of the prevailing environments of deposition and consequent amounts of sediment deposited at different times, and also to the distribution and timing of tectonic events through the state's history that uplifted certain sedimentary sequences and made them susceptible to erosional forces. These factors have made the distribution, thickness, and relative abundance of rocks and fossils of different geologic ages very unequal within the state. The maps accompanying the discussion of each period, as well as the general geologic map of New Mexico, are worth perusing to get a feel for the extent to which various periods are represented in New Mexico.

Table 2 summarizes the sedimentary rock sequence exposed in New Mexico, allowing the relative abundance of rocks of different ages, and the diversity of fossils within them, to be compared. The rocks of some periods, such as the Pennsylvanian and Cretaceous, have yielded more than a thousand different kinds of fossils, whereas fossils from other periods, like the Cambrian, Silurian and Jurassic, are not well represented in the state. The

Table 2. Distribution and extent of rocks and fossils in New Mexico. Data from Northrop (1962), the 1:500,000 geologic map of New Mexico (1965), and the author's unpublished work.

Age	Percentage of northern N.M. area	Percentage of southern N.M. area	Total percentage, entire state	Maximum composite thickness (in meters)	Total number of different fossil species reported
Quaternary	13.2	39.5	26.1	300	75*
Miocene-Pliocene	11.6	13.0	12.3		125
Eocene-Oligocene	5.0	.7	2.9	4,600	200
Paleocene	4.2	.1	2.1		350
Cretaceous	27.4	1.2	14.5	8,350	1,400
Jurassic	3.8	—	1.9	380	10
Triassic	11.0	1.7	6.4	700	100
Permian	6.3	23.5	14.7	3,700	900
Pennsylvanian	3.0	1.2	2.1	2,450	1,250
Mississippian	.1	.1	.1	715	900
Devonian	—	.1	<.1	700	250
Silurian	—	<.1	<.1	425	75
Ordovician	—	.2	.1	730	525
Cambrian	—	<.1	<.1	60	25
Precambrian	.3	.9	2.0	untabulated	—
Igneous/volcanic (all ages)	11.3	17.7	14.5	untabulated	—

*Includes only fossils of extinct species

A piece of a stromatolite, showing fine, undulating laminae. The dark layers contain high concentrations of organic matter.

number of different fossils from rocks of a certain period is partially related to the extent and thickness of the outcrops of that period, but also is affected by the types of environments represented, the quality of preservation, and the degree to which the fossils of a given period have been studied. Also, younger rocks tend to be more widely exposed than older ones, simply because there has been less time for erosion to remove them or for them to be covered by later deposits.

Precambrian

Before the 1950s, few undoubted fossils had been reported from Precambrian rocks anywhere in the world, and we knew virtually nothing about the nature of life during the first five-sixths of earth history. Now, many dozens of fossils are known from this earliest unit of geologic time, some coming from rocks more than 3 billion years old. Precambrian fossils are of two main types—primitive single-celled algae and bacteria found mainly in cherts, and stromatolites. Stromatolites are finely laminated sedimentary structures formed by algal mats, which trap sediments, are covered, and then become reestablished on top of the sediments again. The result is usually a horizontally undulating layered sequence representing dozens of successive algal mats, but sometimes the mats formed thick knobby columns up to a meter in height. We know exactly how Precambrian stromatolites formed because modern stromatolites are forming today in the inter- and supratidal zones of tropical oceans.

Precambrian rocks are widespread in New Mexico, but are generally of volcanic or metamorphic origin, rather than the result of nearshore sedimentary deposition, in which fossils might have been preserved. For this reason, no fossils have yet been reported from the New Mexico Precambrian, but in some areas the right kind of layered rocks seems to be present, so it may only be a matter of time and detailed searching before the earliest remains of life are recognized in our state. Stromatolites have been found in Precambrian rocks in the Franklin Mountains, north of El Paso, Texas, close to the Texas–New Mexico boundary.

Cambrian

The Cambrian Period (500–570 million years ago) is the first to have fossils of complex, shelled, multicellular organisms, and the boundary between the Precambrian and Cambrian has traditionally been placed at the first appearance of shelled metazoans. Trilobites represent about half of all Cambrian fossils, with inarticulate and primitive articulate brachiopods, primitive echinoderms, sponges, molluscs, and the extinct phylum Archaeocyathida also forming parts of Cambrian assemblages. No terrestrial or fresh-water Cambrian fossils are known. The Cambrian represents the first great diversification of animals, and was a time of experimentation in various body plans. In contrast to later periods, the Cambrian seafloor was populated with a relatively low diversity of invertebrates, because several major groups had not yet appeared. A few years ago, the first Cambrian vertebrate fossils were discovered, consisting of a few isolated plates of agnathans.

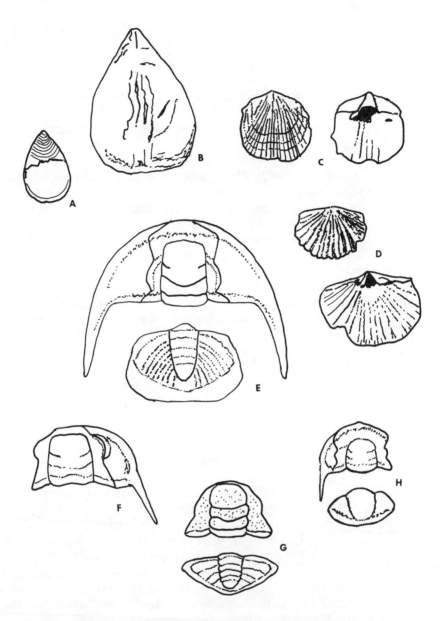

Cambrian fossils. All are of Late Cambrian age. **Inarticulate Brachiopods:** A, *Lingulella*, x 3; B, *Westonia*, x 4. **Articulate Brachiopods:** C, *Billingsella*, showing outside (left) and inside (right) of shell, x 1; D, *Eoorthis*, showing outside (left) and inside (right) of shell, x 1. **Trilobites:** E, *Prosaukia*, cephalon (top) and pygidium (bottom), x 1; F, *Chariocephalus*, cephalon, x 3; G, *Ptychaspis*, cephalon and pygidium, x .75; H, *Camaraspis*, cephalon and pygidium, x 1.

During the Cambrian, shallow seas progressively advanced over what was to become the North American continent; this great transgression climaxed in the Early Ordovician, with all but a few areas in central North America covered by the sea. Cambrian rocks in New Mexico, however, are limited to a few outcrops in the San Andres, Sacramento, Mud Springs, Organ, Franklin, Big Hatchet, Caballo, and Mimbres mountains, the Cooks Range, and isolated exposures in the Silver City area. These Cambrian strata (Bliss Sandstone or Tonuco Formation) are generally thin dark sandstones that have sparse fragmentary fossils; New Mexico's Cambrian fossil record is by far the poorest Paleozoic life record in the state. The most common types are several trilobites, inarticulate brachiopods, and a few primitive articulate brachiopods. These few fossils indicate that only Late Cambrian time is represented in New Mexico; presumably earlier Cambrian sediments were never deposited, or were deposited but have long since been eroded away.

The distribution of Cambrian rocks in the state generally follows that of the much thicker Ordovician sequence.

Ordovician

In the Ordovician Period, 440–500 million years ago, marine invertebrates continued to diversify, and by the end of the period all of the major phyla present today had evolved. New groups entering the fossil record were bryozoans, rugose and tabulate corals, crinoids and other echinoderms, and foraminifers. Trilobites continued to be very common, but less dominant than they had been in the Cambrian, simply because they shared the seafloor with so many other types of invertebrates. Articulate brachiopods became abundant for the first time, attaining a success that would endure until the very end of the Paleozoic. Nautiloid cephalopods, which had appeared in the Late Cambrian, rose to prominence as well, and are one of the most common Ordovician groups. Pelecypods and gastropods, though normally minor elements of Ordovician faunas, were much more successful than they had been in the Cambrian; the same trend characterized the graptolites, which evolved rapidly from forms with many branches to numerous advanced types with only one or two. The bony plates of agnathan vertebrates are present in a few Ordovician deposits around the world, (though none have been reported from New Mexico) but these primitive fish were not yet numerous in the oceans. The

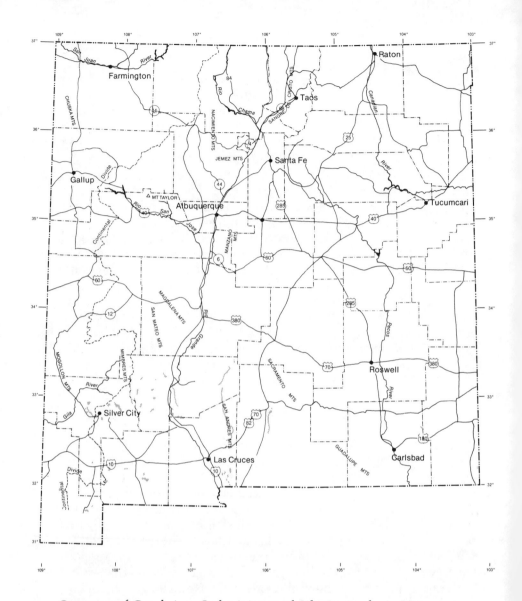

Outcrops of Cambrian, Ordovician, and Silurian rocks in New Mexico.

Reconstruction of a Late Ordovician shallow marine environment as it might have looked in southern New Mexico about 450 million years ago. Straight-coned Nautiloid cephalopods dominate the scene, with colonial tabulate coral heads (lower left), solitary rugose corals (middle left), a large trilobite (center foreground), a few crinoids (center, and towards background), branching bryozoans (left and right foreground), and assorted brachiopods, pelecypods, and gastropods. *(Courtesy of the Smithsonian Institution)*

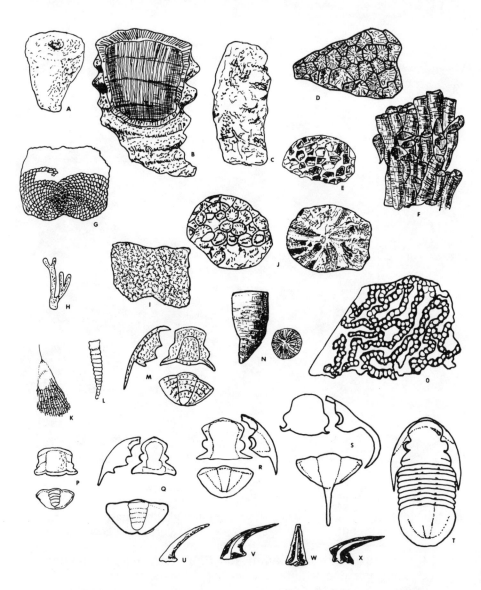

Ordovician fossils. **Sponges:** A, *Calathium*, x 1; B, *Archaeoscypha*, x .2. **Stromatoporoid:** C, *Labechia*, x 1. **Rugose Corals:** D, *Favistina*, x 1; F, *Paleophyllum*, x .75; N, *Streptelasma*, x .5. **Tabulate Corals:** E, *Paleofavosites*, x 1; I, *Protrochistolithus*, x 1; J, *Calapoecia*, x 1; O, *Manipora*, x .5. **Calcareous Algae:** G, *Receptaculites*, x .5. **Bryozoa:** H, *Bythopora*, x 1. **Graptolite:** K, *Dictyonema*, x 1. **Worm (?) tube:** L, *Cornulites*, x 2. **Trilobites:** M, *Hystricurus*, x 1; P, *Leiostegium*, x .75; Q, *Aulacoparia*, x 1; R, *Bellefontia*, x .75; S, *Symphysurina*, x 1; T, *Isoteloides*, x .75. **Conodonts** (all x 10–13): U, *Drepanodus*; V, *Paltodus*; W, *Acontiodus*; X, *Oistodus*.

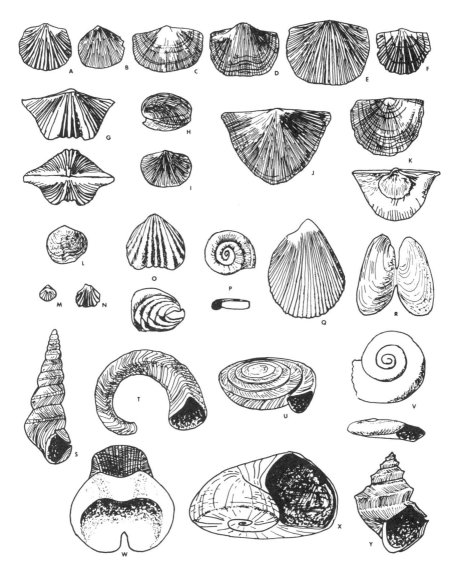

Ordovician fossils. **Articulate Brachiopods:** A, *Nanorthis*, x 1.25; B, *Diparelasma*, x 1; C, *Finkelnburgia*, x 1; D, H, *Hebertella*, top and side views, x 1; E, *Plaesiomys*, x 1; F, *Apheorthis*, x 1; G, *Platystrophia*, top and back views, x .75; I, *Onniella*, x 1; J, *Rafinesquina*, x .75; K, *Strophomena*, exterior and interior shell views, x .75; L, *Diaphelasma*, x 1; M, *Zygospira*, x .75; N, *Hypsiptycha*, x .75; O, *Lepidocyclus*, top and side views, x .75.
Gastropods: P, *Bridgeina*, x .75; S, *Hormotoma*, x 1; T, *Lytospira*, x .75; U, *Ophileta*, x .75; V, *Proliospira*, x 1; W, *Bucania*, x 1; X, *Malcurites*, x .5; Y, *Lophospira*, x 1. **Pelecypods:** Q, *Ambonychia*, x .75; R, *Modiolopsis*, x .75.

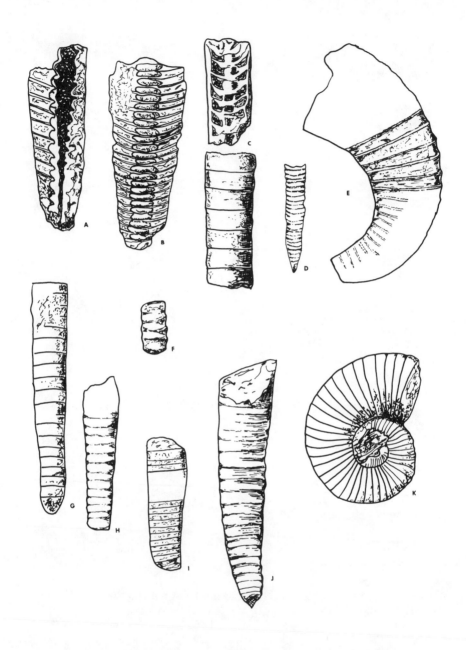

Ordovician fossils. **Nautiloid cephalopods:** A, *Actinoceras*, interior view, x .25; B, *Armenoceras*, interior view, x .25; C, *Buttsoceras*, interior and exterior views, x 1; D, *Clitendoceras*, x .25; E, *Bisonoceras*, x .5; F, *Kyminoceras*, x 1; G, *Michelinoceras*, x 1; H, *Rudolfoceras*, x 1; I, *Rioceras*, x 1; J, *Endocycloceras*, x .5; K, *Campbelloceras*, x .75.

ORDOVICIAN

Ordovician can be viewed as a time of great flourishing of marine faunas, the culmination of the expansion of marine invertebrates that began in the Cambrian.

During Ordovician time, especially in the Early Ordovician, virtually the entire North American continent was covered by warm, shallow seas in which limy sediments were deposited. Extensive mountain-building in the latter part of the period produced an elongate narrow continent along what is now the east coast of North America—the first ancestor of the Appalachian Mountains. In New Mexico thick Ordovician sequences are present in most of the mountain ranges of the south-central and southwestern part of the state, but no Ordovician rocks are exposed north of the San Andres Mountains. Both Early (El Paso Group) and Late (Montoya Group) Ordovician strata are locally profusely fossiliferous, though the preservation of many fossils in massive, sometimes dolomitized limestones makes good specimens difficult to obtain. The fossils of the New Mexico Ordovician are among the best known in the state, due largely to the efforts of Dr. Rousseau Flower, emeritus senior paleontologist at the New Mexico Bureau of Mines and Mineral Resources.

Brachiopods and nautiloids are the most abundant and varied Ordovician fossils in New Mexico. Some inarticulate brachiopods similar to those in the Cambrian are present, but articulates now are far more common. Most types of articulates are orthids, relatively simple forms with wide hingelines, generally flat shapes, and ornamentation consisting of fine radiating ridges, though *Platystrophia* was somewhat wing-shaped, with high ridges that made it resemble some of the spiriferid brachiopods that were common in the middle and late Paleozoic. The nautiloids are primarily straight cones that are rarely completely or excellently preserved. Many of these nautiloids look rather similar on the exterior; accurate identification of many genera depends on detailed examination of the internal shell features. Some New Mexico nautiloids were quite large; a fragment found in the Caballo Mountains indicated a total shell length of 7 to 10 meters! Gastropods and fragmented trilobites are moderately common, and small numbers of sponges, bryozoans, pelecypods, crinoids, and graptolites are also occasionally found in New Mexico Ordovician rocks. A good variety of mostly colonial corals is also present in the Late Ordovician Montoya Group. One of the most bizarre of the Ordovician fossils is *Receptaculites*, rather large disc-shaped or round organisms with numerous small plates that early workers thought were sponges, but that are now considered the skeletons of calcareous algae. Some Ordovician beds show stromatolitic laminae and even large lens-shaped reeflike structures some 125 meters in diameter, testifying to the presence of algal mats in the shallow Ordovician seas of New Mexico.

Some Ordovician fossils remain enigmas. In the early 1960s, Flower described several different kinds of organic objects attached to colonial corals, but no one has been able to figure out exactly what group of animals or plants these fossils belong to. One of them, which resembled a small wart, Flower named *Kruschevia*, after "a certain international figure whose activities in Washington made me seriously late in arriving at the United States National Museum." Let it not be said that paleontologists lack a sense of humor!

Silurian

The Silurian Period, about 400–440 million years ago, was a time of prolific invertebrate marine life. Many of the groups that attained success in the Ordovician diversified and became even more important in the Silurian, especially articulate brachiopods, colonial corals, stromatoporoids, and echinoderms. For the first time, gigantic coral reefs developed, spreading over most of the area from Wisconsin to New York, and associated with these complex organic structures were great numbers of other organisms. The first jawed fish, acanthodians, first appeared in the Silurian, but vertebrates were still minor elements of marine faunas. Trilobites, nautiloids, and graptolites, abundant in the Ordovician, became less so in the Silurian and began evolutionary declines from which none of these groups recovered. An event of great importance occurred in Silurian time, the initial colonization of the land by both animals and plants. The animals were scorpionlike arthropods, and the plants were very primitive vascular forms without leaves or true roots. Initially these plants probably lived only in moist areas along the edges of streams, ponds, and marine shorelines, but as adaptations for a terrestrial existence improved, they expanded over the land and within the next 100 million years their descendants clothed it in vast forests.

Shallow Silurian seas are thought to have covered much of North America, but Silurian rocks are not common in the west-central and southwestern parts of the continent. New Mexico's Silurian rocks are limited to a thin formation, the Fusselman Do-

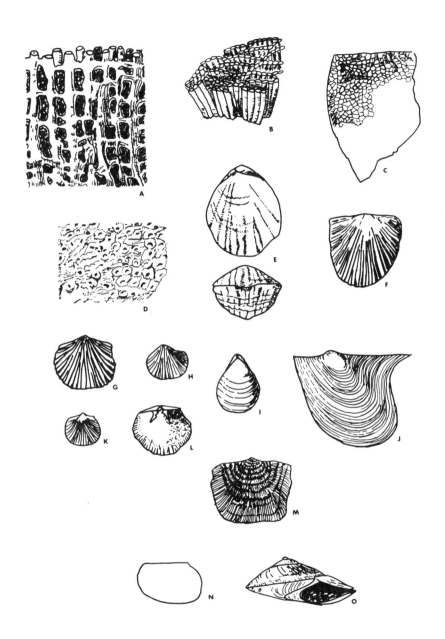

Silurian fossils. **Tabulate Corals:** A, *Syringopora*, x 1; B, *Halysites*, x 1; C, *Favosites*, x 1; D, *Heliolites*, x 1. **Brachiopods:** E, *Virgiana*, top and front views, x 1; F, *Strophodonta*, x 1; G, *Schizoramma*, x .75; H, *Plectatrypa*, x .75; I, *Whitfieldella*, x .75; K, L, *Dalmanella*, top view, x .75; and interior of shell, x 1.5; M, *Leptaena*, x 1. **Pelecypod:** J, *Pterinea*, x .75. **Ostracod:** N, Leperditiid, x 2. **Gastropod:** O, *Liospira*, x 2.

lomite, a brownish-gray massive dolomitic limestone present in limited exposures in most of the mountain ranges of the southern part of the state. Fossils are uncommon and poorly preserved, and, therefore, most identifications have been tentative. The main fossils are brachiopods and corals, indicating an Early to Middle Silurian age for the Fusselman. Some fossil genera carry over from the Ordovician, such as *Dalmanella, Platystrophia, Streptelasma,* and *Hormotoma.* A particularly characteristic Silurian fossil in New Mexico and elsewhere in North America is *Virgiana,* one of a group of pentamerid brachiopods (others include *Conchidium* and *Pentamerus*) that attained fairly large size, ovoid shape, and subdued shell ornamentation; they are found almost wherever Silurian marine deposits are exposed. *Halysites,* a colonial tabulate coral, is commonly called the "chain coral" because of the curious chainlike arrangement of its polyp compartments.

Devonian

The Devonian Period, about 350–400 million years ago, followed the Silurian with few major changes in the dominant marine invertebrates. Articulate brachiopods, especially the wing-shaped spiriferids and related forms, continued their dominance of shallow marine environments, along with stalked echinoderms, bryozoans, and corals. The large reefs established in the Silurian persisted with only minor faunal changes into the Devonian. The major group of graptolites, those having one or two branches, became extinct during the Devonian, and nautiloid cephalopods and trilobites dwindled to become uncommon elements of most marine faunas. Pelecypods and gastropods, on the other hand, while still not abundant, diversified significantly and became more important. Terrestrial plants became more varied and abundant, and by the end of the period the first true forests of sizeable trees developed.

Three outstanding evolutionary events occurred during Devonian time. The first ammonoids, coiled shelled cephalopods that were to become extraordinarily successful in the later Paleozoic and especially in the Mesozoic, arose in the Devonian from nautiloids. These early ammonoids, like *Manticoceras,* were small, had relatively simple sutures, and were not abundant. Second, vertebrates became important and diverse for the first time. The

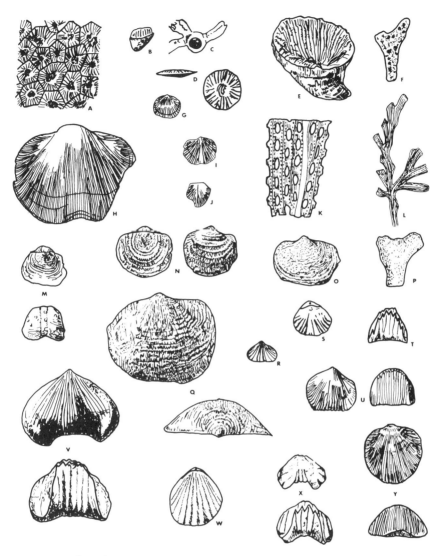

Devonian fossils. **Rugose Corals:** A, *Hexagonaria*, x 1; B, *Macgeea*, x 1; D, *Microcyclus*, x 1; E, *Zaphrenthis*, x .75. **Tabulate Coral:** C, *Aulocaulis*, x 2. **Bryozoa:** F, *Eridotrypella*, x 1; K, *Ptiloporella*, x 3; L, *Sulcoretepora*, x 1; P, *Leptotrypella*, x 1. **Brachiopods:** G, *Rhipidomella*, x .75; H, *Schizophoria*, x 1; I, *Cariniferella*, x 1; J, *Douvillinaria*, x 1; M, *Leioproductus*, top and front views, x .75; N, *Laminatia*, top and bottom views, x 1.5; O, *Sentosia*, x .75; Q, *Planoproductus*, bottom and end views, x .75; R, *Camarotoechia*, x .75; S, *Hyborhynchella*, x .75; T, *Pugnoides*, x .75; U, *Hypothyridina*, top and front views, x .75; V, *Paurorhyncha*, bottom and front views, x 1; W, *Leiorhynchus*, x .75; X, *Porostichtia*, bottom and front views, x .75; Y, *Atrypa*, top and front views, x .75.

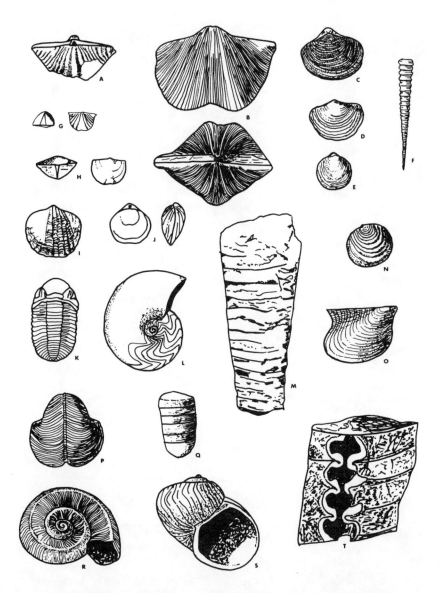

Devonian fossils. **Brachiopods:** A, *Strophopleura*, x 1.5; B, *Cyrtospirifer*, top and end views, x .75; C, *Torynifer*, x .75; D, E, *Cleiothyridina*, two different species, x .75; G, *Cyrtina*, end and top views, x .75; H, *Thomasaria*, end and top views, x .75; I, *Spinatrypa*, x .75; J, *Composita*, top and side views, x .75. **Problematical Mollusc:** F, *Tentaculites*, x .75. **Trilobite:** K, *Phacops*, x .75. **Ammonoid:** L, *Manticoceras*, x .3. **Nautiloids:** M, T, *Macroloxoceras*, exterior, x .2, and interior, x .75, views; Q, *Michelinoceras*, x .75. **Pelecypods:** N, *Paracyclas*, x .75; O, *Actinopteria*, x .75. **Gastropods:** P, *Bellerophon*, x 2; R, *Euomphalus*, x .75; S, *Platyostoma*, x .75.

primitive, heavily armored agnathans, spiny acanthodians, and armored placoderms were by far the most numerous vertebrates, but the first sharks, and ray-finned and lobe-finned bony fish were also present. At the end of the Devonian, the primitive groups dwindled rapidly or became extinct, while the sharks and bony fish expanded to replace them. The Devonian explosion in aquatic vertebrates has led some paleontologists to refer to the Devonian as the "Age of Fishes." Finally, near the end of the Devonian, the first terrestrial vertebrate, a primitive amphibian, evolved from a group of lobe-finned fish, thus establishing vertebrates on land and setting the stage for their subsequent rapid expansion into this new habitat.

Across North America the seas had retreated at the end of the Silurian, leaving much of the continent above sea level, but in the middle and late Devonian extensive transgressions across North America began again, leaving marine sediments in New Mexico and most of the other western states. In the east, renewed mountain-building occurred as the ancestral North American continent collided with the western European landmass, closing the proto–Atlantic Ocean that had existed in earlier Paleozoic time.

Relatively thick sequences of Middle to Late Devonian rocks, mostly dark shales, are present in most of the mountain ranges of south-central and southwestern New Mexico. The most fossiliferous deposits are in the Silver City–Hillsboro–Lake Valley area, and in the San Andres and Sacramento mountains, and these strata have yielded a wide variety of fossils. New Mexico Devonian fossils are generally better preserved than fossils of previous periods, and in some localities, like Sly Gap in the San Andres mountains, thousands of specimens erode out of the shaly rocks. The majority of these fossils are brachiopods, including many species and some genera that were first described from New Mexico. For the first time strophomenids, the group that includes the large and highly convex productoids (e.g., *Planoproductus, Laminatia,* etc.), and the wing-shaped spiriferids (e.g., *Cyrtospirifer*) are abundant, but most other important brachiopod groups are represented as well. Corals and bryozoans are moderately common, and crinoid stem segments are abundant in some beds. Gastropods, pelecypods, nautiloids, ammonoids, trilobites, and conodonts are found only occasionally. A few spines and scales of sharks and placoderms represent New Mexico's earliest vertebrate record.

THE NEW MEXICO FOSSIL RECORD

Mississippian

At the end of the Devonian Period, most marine environments in the United States were subjected to the deposition of black shales that are sparsely fossiliferous. Although the explanation for this widespread depositional pattern is not clear, it apparently had severe consequences for marine invertebrates, resulting in the extinction of many groups that had been common in the Silurian and Devonian. As the Mississippian Period (310–350 million years ago) opened, the composition of marine faunas was significantly different from earlier in the Paleozoic. The gigantic coral-stromatoporoid reefs of the Silurian and Devonian were gone; Mississippian reefs, including some in New Mexico, were restricted to small lens-shaped mounds on the seafloor. Crinoids and blastoids reached their greatest success in the Mississippian, growing in profusion over large areas of seafloor, and some Mississippian limestones are composed entirely of millions of their plates and ossicles. Brachiopods continued to be extremely abundant and diverse; two-thirds to three-fourths of all Mississippian brachiopods were strophomenids and spiriferids. Bryozoans were also successful, with fenestrate types predominating. A curious and distinctive Mississippian bryozoan is *Archimedes*, which had a central screwlike axis from which the fenestrate colony projected in elaborate spiral sheets. Foraminifers are much better represented in Mississippian than in older rocks; the most common and distinctive group are the endothyrids, which are present in vast numbers in some limestones. Ammonoids were more abundant than in the Devonian, but are still relatively uncommon fossils, and nautiloids and trilobites are unusual finds in Mississippian rocks. Pelecypods, gastropods, and corals, though successful, were generally subsidiary members of marine assemblages. Graptolites became extinct. Among the vertebrates, sharks and ray-finned bony fishes dominated, with only remnants of the more primitive heavily armored fish groups surviving into the Mississippian. On land, forests of lower vascular plants spread and diversified; within them lived arthropods and amphibians.

New Mexico Mississippian rocks are predominantly limestones and are more widespread than the strata of any earlier geologic period. Thick, highly fossiliferous sequences are present in the San Andres, Sacramento, and Magdalena mountains, near the town of Lake Valley, and in southwestern New Mexico, particularly in the Big Hatchet Mountains and several other smaller mountain ranges. In addition, Mississippian rocks were discov-

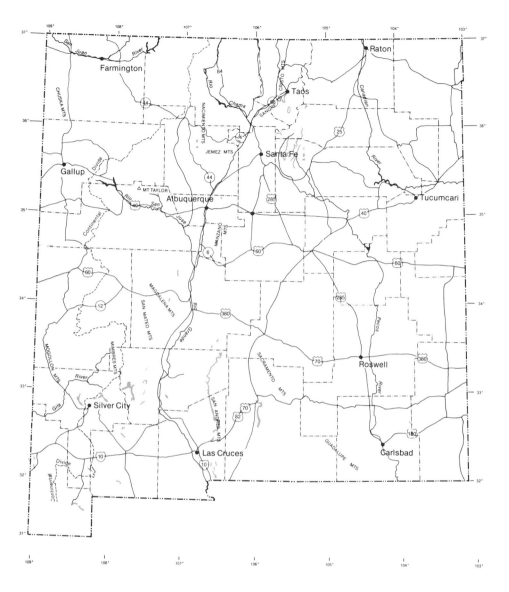

Outcrops of Devonian and Mississippian rocks in New Mexico.

Mississippian shallow marine environment, showing a dense crinoid "garden" and starfish, as it might have looked in the Lake Valley area some 330 million years ago. *(Courtesy of the Smithsonian Institution)*

ered in northern New Mexico in the mid 1950s, and relatively limited exposures are now known in the Sandia Mountains, Jemez area, and several places in the Sangre de Cristo Mountains along the Pecos River and north and northwest of Las Vegas. Most of the recent paleontological work on the Mississippian in New Mexico, including the first report of Mississippian fossils in northern New Mexico, has been done by Dr. A. K. Armstrong, of the United States Geological Survey.

New Mexico Mississippian fossils follow the same patterns shown in other parts of the United States. Several types of fossils, such as *Archimedes*, *Endothyra* and related foraminifers, and blastoid echinoderms, when found are sufficient to identify immediately a given outcrop as Mississippian. A wide variety of brachiopods are also distinctive of the New Mexico Mississippian, with a few genera such as *Schizophoria* and *Rhipidomella* continuing into the Mississippian from the Devonian. Corals, mainly solitary conical rugose corals, are common locally. Thin sections showing the intricate septa and other internal shell structures are generally required for accurate identification of these corals, which all look fairly similar externally. Crinoid stem segments are abundant, as they are in most post-Cambrian Paleozoic rocks, but in addition New Mexico has several concentrations of exquisitely preserved complete crinoid calyces. Complete calyces are generally rare as fossils, because their plates disarticulate rapidly after death, or, as one expert suggested, because Paleozoic sharks developed a preference for crinoids and swam around biting the calyces off stems and chewing them up. One of the most famous crinoid localities in the world is near the semighost town of Lake Valley, where literally dozens of different species are represented by complete calyces.

Pennsylvanian

Pennsylvanian (280–310 million years ago) fossils display many continuations and elaborations of trends begun in the Mississippian. Productoid and other strophomenid and spiriferid brachiopods dominated shallow marine communities; rugose corals and crinoids are also conspicuous Pennsylvanian fossils. Pelecypods, gastropods, and ammonoids are all more common and diverse than at any previous time, whereas nautiloids, trilobites, tabulate corals, and eurypterids continue their evolutionary decline and are only occasionally found. For the first time in the Pennsyl-

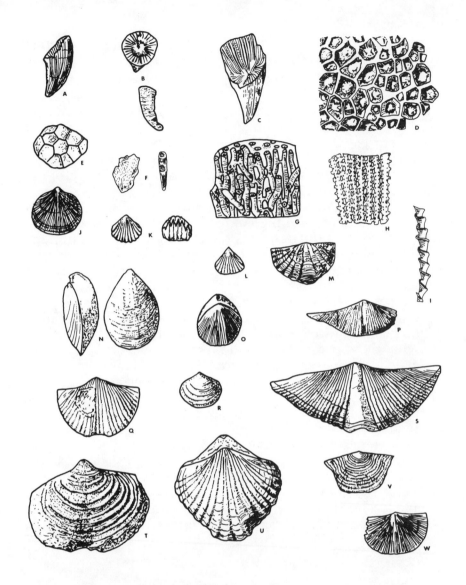

Mississippian fossils. **Rugose Corals:** A, *Rylstonia*, x .75; B, *Homalophyllites*, top and side views, x .75; C, *Amplexizaphrentis*, x .75; D, *Lithostrotionella*, x .75. **Tabulate Corals:** E, *Cleistopora*, x .75; F, *Palaecis*, side and top views, x .75; G, *Syringopora*, x .75. **Bryozoa:** H, *Fenestella*, x 4; I, *Archimedes*, spiral axis, which supported *Fenestella*-like colony, x 1. **Brachiopods:** J, *Rhipidomella*, x .75; K, *Rhynchopora*, top and front views, x .75; L, *Tetracamera*, x .75; M, *Tylothyris*, x 1.25; N, *Beecheria*, side and bottom views, x .75; O, *Eumetria*, x .75; P, *Unispirifer*, x .75; Q, *Spirifer*, x .75; R, *Cleiothyridina*, x .75; S, *Syringothyris*, x .5; T, *Athryis*, x .75; U, *Brachythyris*, x .75; V, *Leptaena*, x .5; W, *Chonetes*, x 1.25.

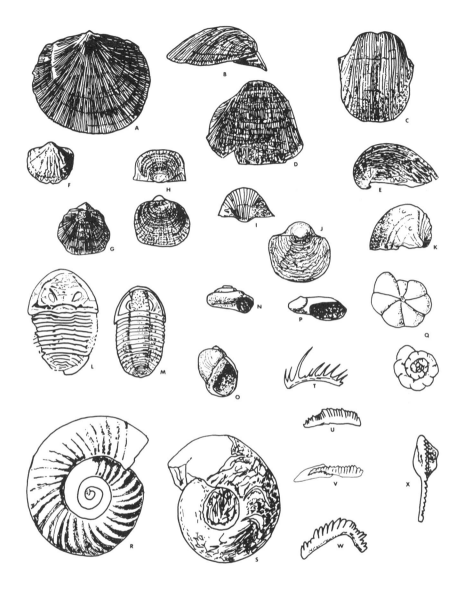

Mississippian fossils. **Brachiopods:** A, B, *Orthotetes*, top and side views, x 1; C, *Inflatia*, x .75; D, E, *Stegacanthia*, front-bottom and side views, x .75; F, *Productella*, x .75; G, *Ovatia*, x .75; H, *Marginatia*, end and top views, x .75; I, J, *Productina*, end and top views, x 1.25; K, *Geniculifera*, x 1.25. **Trilobites:** L, *Proetus*, x 1; M, *Breviphillipsia*, x 1. **Gastropods:** N, *Straparollus*, x .75; O, *Naticopsis*, x .75; P, *Platyceras*, x .75. **Foraminifer:** Q, *Endothyra*, cross-sectional and external views, x 12. **Ammonoids:** R, *Pericyclus*, x 1; S, *Gattendorfia*, x 1. **Conodonts** (all x 10–12): T, *Hindeodella*, U, *Spathognathodus*, V, *Polygnathus*, W, *Bryantodus*, X, *Gnathodus*.

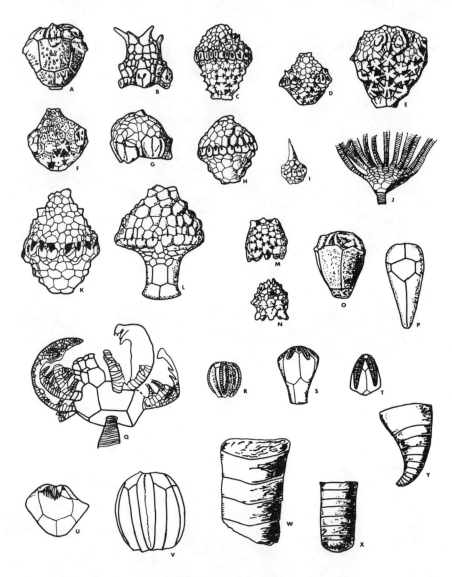

Mississippian fossils. **Crinoids** (most are calyces, without the arms): A, *Platycrinites*, x .75; B, *Amphoracrinus*, x .75; C, *Cactocrinus*, x .75; D, *Actinocrinites*, x .75; E, *Physetocrinus*, x .75; F, *Steganocrinus*, x .75; G, *Agaricocrinus*, x .75; H, *Batocrinus*, x .75; I, *Dorycrinus*, x .75; J, *Eretmocrinus*, x .75; K, *Nunnacrinus*, x 1; L, *Uperocrinus*, x .75; M, N, *Rhodocrinites*, two different species, x .75; O, *Kallimorphocrinus*, x 15; P, *Passelocrinus*, x 25; Q, *Eucladocrinus*, x .75. **Blastoids:** R, *Cryptoblastus*, x 1; S, *Orophocrinus*, x 1; T, *Pentremites*, x .75; U, *Hadroblastus*, x 1; V, *Monadoblastus*, x 4.
Nautiloids: W, *Hesperoceras*, x 1; X, *Mooreoceras*, x 1; Y, *Welleroceras*, x .25.

PENNSYLVANIAN

Large numbers of rice-sized fusulinids are found in some Pennsylvanian limestones. The numerous football-shaped fusulinids in this photo are slightly larger than life-sized. From Jemez Springs area.

vanian, echinoids become relatively abundant, and isolated spines and plates are fairly common in some limestones. A change of greater importance is the proliferation of the football-shaped foraminifers called fusulinids into one of the most abundant and characteristic fossils found in the Pennsylvanian. These rice-sized fossils evolved from endothyrid forams at the end of the Mississippian; they started out small and discoid (e.g., *Millerella*) but evolved through the Pennsylvanian to become large, more complex internally, and more fusiform in shape. In areas adjacent to the shorelines of the shallow seas that covered most of western and central North America during the Pennsylvanian, lush tropical forests grew, filled with tree-sized representatives of several lower vascular plant groups (*Lepidodendron, Calamites,* and various ferns) and some of the earliest seed-bearing gymnosperms (*Cordaites, Walchia*) and seed-ferns (*Neuropteris*). The coal that

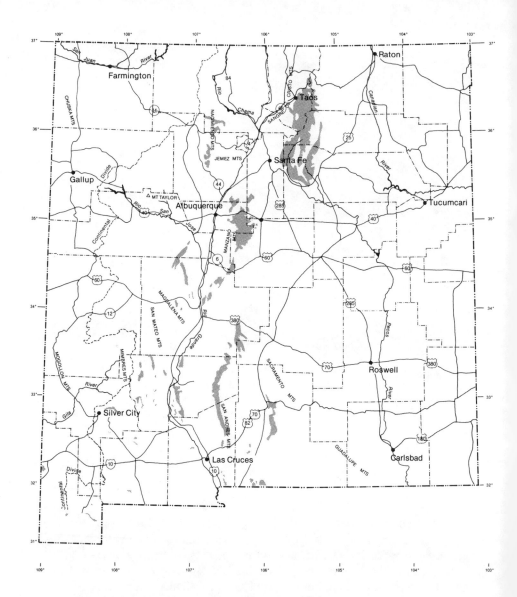

Outcrops of Pennsylvanian rocks in New Mexico.

PENNSYLVANIAN

A branch of *Walchia*, a primitive conifer, showing the needlelike leaves. From a Pennsylvanian shale in the Manzano Mountains.

was formed from these ancient forests has been mined for more than a century in large areas of Kentucky, Ohio, West Virginia, Pennsylvania, Indiana, and Illinois.

In ponds and streams within these forests were large amphibians, a variety of ray-finned and lobe-finned fish, eurypterids, shrimp, ostracods, and a few freshwater pelecypods; insects of many different kinds were abundant in the air above the forested swamps. Also present, though not yet common, were some of the earliest reptiles, generalized lizard-shaped creatures not much different structurally from their amphibian ancestors, but with the critical ability to lay amniotic eggs on land, without the necessity of having them or the newly hatched young develop in water.

By the Pennsylvanian Period the isolated continents of the earlier Paleozoic world had coalesced to form Pangaea, an enormous

THE NEW MEXICO FOSSIL RECORD

A small part of the trunk of *Lepidodendron*, showing diamond-shaped leaf scars. Original is about 50 cm long.

landmass composed of what would later become the continents of Africa, Antarctica, South America, Australia, North America, and part of Eurasia. On the southern part of this great landmass (Gondwanaland), glaciers grew and spread over millions of square miles during the Pennsylvanian and Early Permian, to produce an epoch of glaciation more extensive than the more recent northern Pleistocene glaciations. The changes in climate and sea level accompanying these glaciations probably affected sedimentary deposition considerably, and may have been partly responsible for the remarkably rhythmic packages of successively similar strata called cyclothems that are an outstanding feature of Pennsylvanian and Early Permian sequences in the central United States.

Most of New Mexico was under water at this time, with numerous elongate north-south trending landmasses supplying eroded sediments to shallow shelf areas. The largest of these giant

PENNSYLVANIAN

Front portion of *Adelophthalmus*, one of the last genera of eurypterids. Note the several appendages projecting from beneath the somewhat triangular head, including a pair of elongate paddles near the end of the head. Part of a *Cordaites* leaf is visible in the upper part of the photo. In the same sequence of beds at this locality have come many types of plants, and literally dozens of different kinds of Pennsylvanian insects. From the Lucero Mesa area, west of Los Lunas.

New Mexico islands were situated northwest of Santa Fe (extending into Colorado), south of Raton, around the Zuni Mountains, and from around Roswell extending nearly to Santa Fe. Exposures of Pennsylvanian rocks in New Mexico have a much wider geographic distribution than those of any previous period, and marine sequences hundreds of meters thick are found in most parts of the state. Virtually wherever Pennsylvanian rocks are present fossils may be found, and there are numerous places where marine invertebrates erode from the rocks by the thousands. In the Sandia and Manzano mountains east of Albuquerque; along San Diego Canyon between Jemez Pueblo and Battleship Rock and northward into the Nacimiento Mountains; in the Sangre de Cristos near Santa Fe and along the Pecos River north to the area around Taos and Eagle Nest; and east of Alamogordo and Tularosa in the Sacramento Mountains are all localities where well-preserved, highly diverse Pennsylvanian invertebrate fossils are abundant and easy to collect.

THE NEW MEXICO FOSSIL RECORD

Pennsylvanian shallow marine environment as it might have looked in New Mexico about 300 million years ago. Sponges (middle left and center), solitary rugose corals (middle right), brachiopods *(Derbyia, Neospirifer;* center foreground), a large coiled nautiloid (right foreground), erect pelecypods *(Aviculopecten;* middle center), several types of gastropods, a large echinoid (center foreground), and a crinoid are shown. *(Courtesy of the Smithsonian Institution)*

The most common fossils are brachiopods, particularly productoids (like *Antiquatonia* and *Juresania*), flat strophomenids *(Derbyia)*, wing-shaped spiriferids (such as *Neospirifer* and *Punctospirifer*), and ovoid smooth spiriferids *(Composita)*. Crinoid stems (though unfortunately not many calyces), fusulinids, and fenestrate bryozoans are also abundant. Some strata representing relatively nearshore marine environments have large numbers of gastropods and pelecypods, but these are only locally common; most Pennsylvanian limestones have rather few molluscs. In a few localities in the Nacimiento Mountains, steinkerns of giant discoidally coiled gastropods up to seven cm long are present in large numbers.

Corals are moderately common. Basketball-sized mound-shaped colonies of the colonial tabulate coral *Chaetetes* (eroded specimens are sometimes mistaken for petrified wood), and large sol-

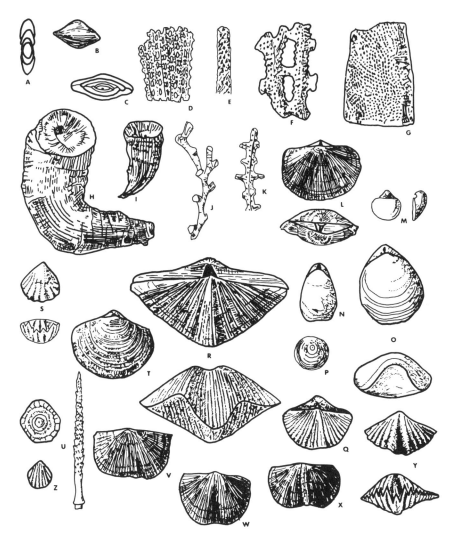

Pennsylvanian fossils. **Fusulinid Foraminifera:** A, *Millerella*, cross-section, x 35; B, *Beedeina*, x 2; C, *Fusulinella*, x 4. **Bryozoa:** D, *Fenestella*, x 5; E, *Rhombopora*, x 5; F, *Polypora*, x 5; G, *Tabulipora*, x 5; K, *Penniretepora*, x 5. **Rugose Corals:** H, *Caninia*, x .75; I, *Lophophyllidium*, x 1. **Tabulate Coral:** J, *Cladochonus*, x 1.
Brachiopods: L, *Derbyia*, x 1; M, *Crurithyris*, x 1; N, *Beecheria*, x 1; O, *Composita*, top and front views, x 1; P, *Orbiculoidea*, an inarticulate brachiopod, x 1; Q, *Anthracospirifer*, x .75; R, *Neospirifer*, top and front views, x 1; S, *Wellerella*, x 1.5; T, *Cleiothyridina*, x 1.5; V, *Neochonetes*, x 1.5; W, *Chonetinella*, x 1.5; X, *Mesolobus*, x 1.5; Y, *Punctospirifer*, bottom and front views, x 1.5; Z, *Hustedia*, x 1.
Echinoids: U, *Archaeocidaris*, single interambulacral plate (left) and single spine (right), x 1.

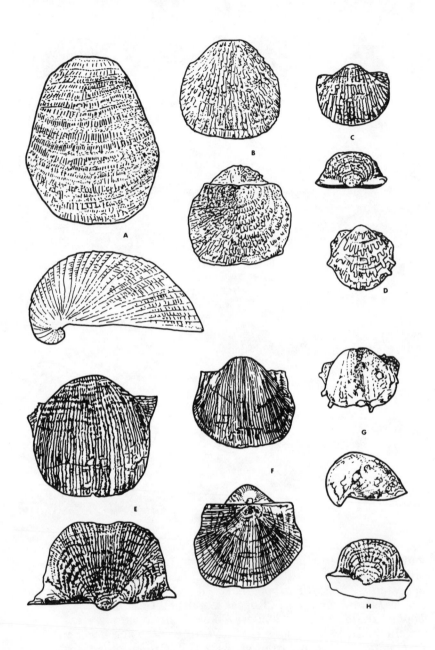

Pennsylvanian fossils. **Brachiopods:** A, *Echinaria*, bottom and side views, x 1; B, *Juresania*, bottom and top views, x 1; C, *Desmoinesia*, bottom and end views, x 1.5; D, *Cancrinella*, x 2; E, *Antiquatonia*, bottom and end views, x 1; F, *Linoproductus*, bottom and top views, x 1; G, *Hystriculina*, bottom and side views, x 1.5; H, *Sandia*, end view, x 1.5.

Pennsylvanian fossils. **Pelecypods:** A, *Phestia*, x 1; B, *Nuculopsis*, x 1; C, *Edmondia*, x 1; D, *Schizodus*, x .75; E, *Wilkingia*, x 1; F, *Pteronites*, x .25; G, *Orthomyalina*, x .75; H, *Aviculopecten*, x 1; I, *Acanthopecten*, x 1; J, *Myalina*, x .75; K, *Astartella*, x .75; L, *Parallelodon*, x 1. **Gastropods:** M, *Stephanozyga*, x 1; N, *Pseudozygopleura*, x 1; O, *Naticopsis (Jedria)*, x 1; P, *Bellerophon*, x 1; Q, *Euphemites*, side and bottom views, x 1; R, *Retispira*, x 1.5; S, *Pharkidonotus*, x 1; T, *Taosia*, x 1; U, *Soleniscus*, x 1; V, *Trachydomia*, x 1; W, *Amphiscapha*, x .75; X, *Anomphalus*, x 5; Y, *Orthonychia*, x 1; Z, *Worthenia*, x 1; AA, *Glabrocingulum*, x 1.5; EE, *Euomphalus*, x 1.

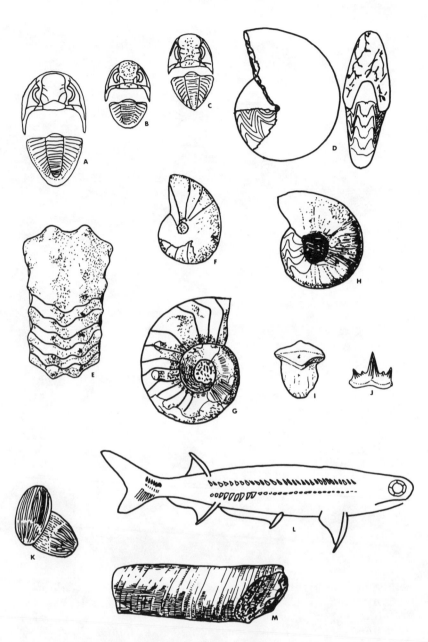

Pennsylvanian fossils. **Trilobites:** A, *Ameura*, x 1; B, *Ditomopyge*, x 1; C, *Anisopyge*, x 1. **Ammonoids:** D, *Gonioloboceras*, side and front views, x 1; H, *Gastrioceras*, x .3. **Nautiloids:** E, *Tainoceras*, top view, x .5; F, *Liroceras*, x 1; G, *Endolobus*, x 1. **Shark teeth:** I, *Petalodus*, x 1; J, *Cladodus*, x .5. **Bony Fish:** K, Coelacanth scales, x 3. **Acanthodian:** L, *Acanthodes*, x .5. **Scaphopod:** M, *Prodentalium*, x 1.

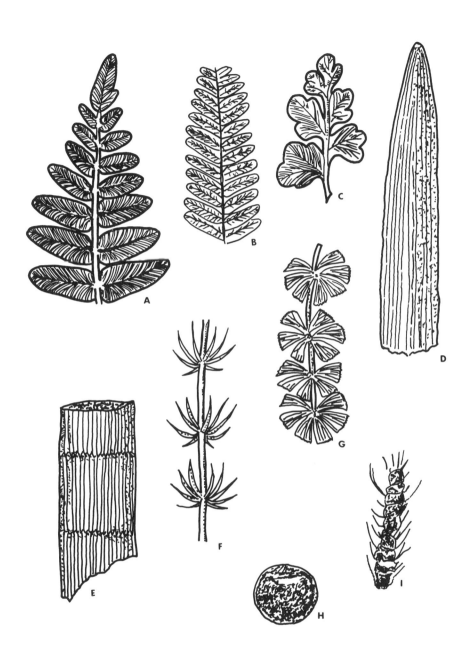

Pennsylvanian fossils. **Plants:** A, *Neuropteris*, a seed fern, x .75; B, *Pecopteris*, a fern, x 1; C, *Sphenopteris*, a seed fern, x 1; D, *Cordaites*, a primitive gymnosperm, x .6; E, *Calamites*, a sphenophyte, x .5; F, *Asterophyllites*, a sphenophyte, x 1; G, *Sphenophyllum*, a sphenophyte, x 1; H, *Cardiocarpus*, a seed from either a seed fern or primitive gymnosperm, x 1; I, *Lepidostrobus*, a lycopod cone, x 1.

Chaetetes, a coral that formed large, mound-shaped colonies. Original is about 25 centimeters high. The colonies consist of thousands of very small polygonal tubes, the compartments where the polyps lived. From the Jemez Springs area.

itary horn corals (like *Caninia*) are conspicuous in some places, particularly in the Jemez area. Coiled nautiloids and ammonoids are rather rare; probably the most common cephalopods are small straight-cone nautiloids. Also present in some Pennsylvanian deposits are extremely large scaphopods, which reached lengths of 20 cm or more, far larger than any modern forms. Pieces of these scaphopods superficially resemble straight-coned nautiloids but may be distinguished by the numerous fine lirae that run the length of the shell.

Plant fossils are occasionally found in large numbers as carbonized impressions in dark, fine-grained swamp or delta deposits, and casts of impressions of tree trunks (*Lepidodendron*, *Calamites*) sometimes turn up in sandstones. Little actual coal, however, was formed from Pennsylvanian plants in New Mexico. In some places, such as Mesa Lucero west of Los Lunas, and in the

Manzanita Mountains east of Albuquerque, insects, crustaceans, eurypterids, amphibians, and fish are preserved in the swamp sediments, giving us an unusual view of life in nonmarine environments during Pennsylvanian time. Together with fossils of the Cretaceous Period, Pennsylvanian fossils are the most abundant, diverse, and widely distributed fossils in the state.

Permian

The Permian Period (230–280 million years ago) began with very little change from the Pennsylvanian, and it is often difficult to distinguish Late Pennsylvanian from Early Permian fossils. The boundary between the two periods is placed at the level where a new fusulinid and a new plant genus appear, but in general both marine and terrestrial fossils on both sides of the boundary are very similar. In New Mexico many of the invertebrate fossils illustrated for the Pennsylvanian continue into the Permian. These include all of the bryozoans and most of the brachiopods, pelecypods, and gastropods. Figures of fossils presented in this section are mainly of genera that are restricted to or are most conspicuous in the Permian of the state.

A major development in the Permian was the growth once again of large reefs. The best and most thoroughly studied example in the world is the great Permian reef complex of southeastern New Mexico and West Texas. Fossils from the reef complex were first collected in 1855, and it is now known that the reefs and associated limestones stretch for hundreds of kilometers in a great arc, with the reefs themselves separating basinal marine deposits to the south from shallow carbonate shelf environments to the north. These reefs, comparable in size to the modern Great Barrier Reef of Australia, were among the most highly diverse and complex associations of organisms the world has ever seen, and they attained mountainous proportions; in fact a good part of the Guadalupe Mountains of New Mexico and Texas consists of these gigantic organic structures.

The reefs were primarily composed of calcareous algae, sponges, and bryozoans; corals were relatively uncommon in them, but vast numbers of other organisms, especially brachiopods, echinoderms and molluscs lived in or around the reefs. Some brachiopods responded to the reef environment by developing bizarre shapes, including some that have the superficial appearance of solitary horn corals. A complete study of the brachiopods of the

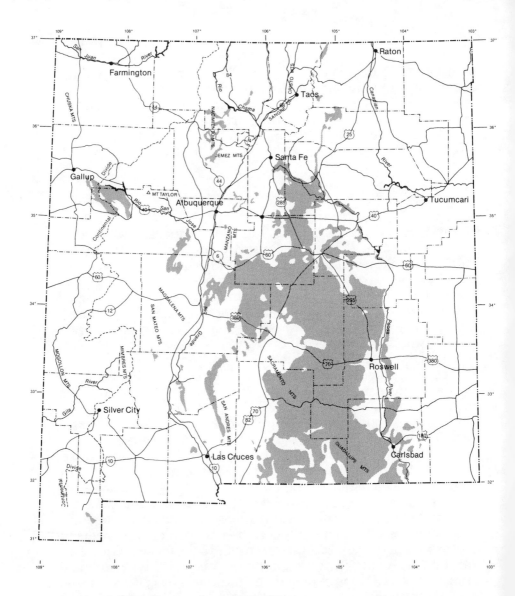

Outcrops of Permian rocks in New Mexico.

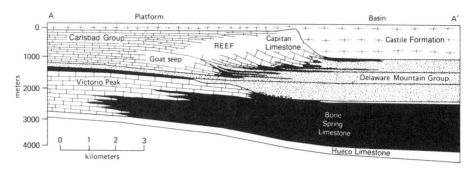

Cross section through the rocks of the Permian reef complex. The Capitán Reef grew on the edge of a shallow cabonate shelf, separating the shelf from the basin to the south. *(After P. B. King)*

reef system was recently finished; it describes hundreds of species and runs an astounding 3,500 pages, which provides an idea of the tremendous diversity and abundance of life in West Texas and southeastern New Mexico during middle and late Permian time. Most of the fossils are silicified, which allows their removal from the dense limestones by acid-etching. Tons of rock have been extracted from the reef complex by paleontologists and literally hundreds of thousands of exquisitely preserved fossils have been etched out of it.

The reef complex grew in shallow marine conditions for more than 20 million years. North of the complex, toward central New Mexico, are extensive marine limestones in which invertebrate fossils are abundant, particularly in the San Andres, Sacramento, and other southern New Mexico mountain ranges. A few Permian marine fossils have also come from restricted exposures as far

THE NEW MEXICO FOSSIL RECORD

Reconstruction of the marine life on a patch reef in the Permian reef complex in southeastern New Mexico and West Texas. Many types of brachiopods are conspicuous on the reef, including the spiny productoids and wing-shaped spiriferids; solitary rugose corals (center of reef) and sponges (far left and far right) are also common. A spiny coiled nautiloid and a group of bizarre brachiopods *(Collemateria,* or *Leptodus)* are seen in the center foreground. *(Courtesy of the Smithsonian Institution)*

north as Bluewater Lake and the Zuni Mountains, northwest of Grants. Many fossils can be easily observed in the Guadalupe Mountains near Carlsbad Caverns. The Permian of northern New Mexico, however, consists mainly of red sandstones that lack marine fossils. The Pennsylvanian sea that had covered most of New Mexico retreated through Permian time, to be replaced by terrestrial environments from north to south, and eventually, near the end of the Permian, the sea left New Mexico completely and the reefs died. In the briny remnants of the sea various minerals precipitated, producing thick salt and potash beds; later, petroleum and gas became trapped in the interstices of the buried reefs. The extraction of these materials from the reef complex contributes significantly to New Mexico's economic well-being. Much later, only a few million years ago, slightly acidic ground water, percolating through the reef complex limestones, created Carlsbad

Caverns. By descending into the caverns one is able literally to walk through the inside of the reef.

The shallow carbonate-shelf limestones of central and southern New Mexico contain profuse marine invertebrate fossils. Many of these are types similar to those that inhabited the reefs, but shelf faunas also show significant differences from the reef faunas. Fusulinid forams are conspicuous in Permian shelf deposits, and in general they show a continuation of trends displayed in the Pennsylvanian. Permian fusulinids are generally more elongate and larger than earlier forms, some reaching widths of several centimeters, which make them among the largest single-celled organisms ever to have existed. An unusual fossil occasionally observed in shelf environments is *Helicoprion*, which represents a spiral whorl of teeth from a medium-sized shark. Instead of the teeth dropping out after they had reached the outer edge of the jaw, as in most sharks, in *Helicoprion* they remained in the jaw and grew downward and inward to form the growing spiral.

In the terrestrial environments of central and northern New Mexico, late Paleozoic forests continued, with plants similar to those of the Pennsylvanian Period, but the climate was drier, and dense, lush forests were not nearly as widespread in the Permian as in the Pennsylvanian. Just south of Jemez Springs, and in the Oscura Mountains, plant fossils were replaced by gray, blue, and green copper minerals, and were mined as part of the copper ore. Fish, amphibians, and increasing numbers of reptiles lived in and around these forests. Vertebrate fossils, including some complete skeletons, have been found in Early Permian red sandstones near Jemez Springs, in El Cobre Canyon near Abiquiu, on Poleo Creek west of Coyote, in the upper Pecos Valley near Galisteo, about 10 miles northeast of Socorro, and in a few other places in New Mexico.

Permian vertebrates were among the earliest vertebrate fossils collected in New Mexico. David Baldwin, a professional fossil collector employed by Othniel C. Marsh, shipped a large amount of material from the Abo Formation to Marsh at Yale University in 1877. Marsh ignored them until, spurred by discoveries made by his great and bitter rival Edward Drinker Cope in Texas, he hurriedly unpacked a few of Baldwin's boxes and quickly wrote a paper describing (some incorrectly, as it turned out) several new genera and species of reptiles and amphibians, in an effort to beat Cope into print. Cope was not amused, and this incident fueled the antagonism between these two great vertebrate paleontolo-

THE NEW MEXICO FOSSIL RECORD

Dimetrodon, the "sail-backed reptile," was in the group that was eventually to lead to the development of mammals. It has been suggested that the function of the "sail" was to aid the animal in soaking up or dissipating heat rapidly, an early experiment in body temperature regulation. *Dimetrodon* grew to a length of nearly three meters; known from the Early Permian of Texas for more than a century, remains of this carnivore were recently discovered in New Mexico, near Jemez Springs (after Charig).

gists. It was not until about 1910 that the rest of Baldwin's Permian collections were examined, and they were found to include a nearly complete skeleton of a previously unknown, very primitive reptile, *Limnoscelis*, which even seventy years later remains the best fossil of that genus ever found. This discovery spurred a 1911 paleontological expedition to the New Mexican Permian, and several others have recovered good and often new reptile and amphibian fossils since then.

Besides *Limnoscelis*, the reptiles include forms like *Dimetrodon* (the "sail-backed" reptile) and *Sphenacodon*, carnivores that reached lengths of 3 meters and were the largest animals living in New Mexico at this time. Another reptile, *Ophiacodon*, was smaller and apparently aquatic, feeding on fish. Amphibians are represented by large bulky animals that resembled stubby 2-meter-long alligators (*Eryops*,), smaller salamanderlike forms,

Reconstruction of *Limnoscelis*, a primitive reptile more than a meter long, that lived in north-central New Mexico during deposition of the red Abo Sandstone (P. S. Lungé).

and members of a group that is structurally intermediate between reptiles and amphibians (*Diadectes*). A few freshwater shark teeth and spines, and bony fish scales have also come from the Permian redbeds, and recently in the Sangre de Cristo Mountains, large ellipsoid trace fossils have been reported that probably were lungfish burrows.

Toward the end of the Permian, the prolific marine faunas exemplified by the Permian reefs began to decline, heralding the onset of the most stunning crisis the life of this planet has ever endured. A massive episode of extinctions at the end of the Permian eradicated from 50 to 90 percent of the species living at the time, and many groups that had been abundant and successful during the Paleozoic became extinct or so depleted that they never regained their former prominence. Most brachiopods, including most of the highly abundant late Paleozoic forms, the main groups

Top: Ophiacodon was an amphibious reptile that probably fed on fish in ponds and streams of Early Permian New Mexico. It measured about two meters in length (P. S. Lungé). Middle: *Eryops* was a large (two-meter-long) bulky Early Permian amphibian. Though it undoubtedly spent a good deal of time in the water, the large and well-developed limbs indicate that it was adept on land as well. This group represents a very successful line of late Paleozoic amphibians that dwindled as the reptiles diversified in the Triassic (after Romer). Bottom: The skeleton of *Sphenacodon* closely resembles that of *Dimetrodon*, except for the lack of a "sail." Complete skeletons have been collected from the Abo Sandstone in the Jemez Springs area. *Sphenacodon* was one of the largest animals of its time, measuring up to three meters in length (after Romer).

Reptile footprints in a slab of Abo Sandstone from Fra Cristobal Mountains. About natural size.

of Paleozoic bryozoans (like the fenestrates), fusulinids, nearly all crinoids and other echinoderms, tabulate and rugose corals, and almost all ammonoids became extinct at the end of the Permian. Also becoming extinct were groups like trilobites and eurypterids, which had been on the decline before the Permian.

The causes of this crisis are still being debated by geologists. Suggestions have ranged from catastrophic events, such as meteor impact or a nearby supernova explosion, to the sea becoming too salty or not salty enough for most invertebrates, to a precipitous sea level drop because of the late Paleozoic glaciations, to the idea that the coalescence of continents that formed Pangaea was responsible. The last idea has much to recommend it; formation of a supercontinent would disrupt climatic and oceanic current patterns, lower sea level, and result in much less continental shelf space (the optimum habitat for marine invertebrates) being available than around numerous smaller but separated continental masses, such as existed throughout most of the Paleozoic. Whatever the correct explanation or explanations, the world that emerged at the beginning of the Triassic Period was a far different one, greatly depleted of many life forms, than that which had existed during the paleontologically glorious time of the late Paleozoic.

Permian fossils. **Fusulinid Foraminifera:** A, *Triticites*, x 2; B, *Dunbarinella*, x 2; C, *Polydiexodina*, x 2. **Sponges:** D, *Polyphymaspongia*, x 1; E, *Amblysiphonella*, x .5; F, *Girtyocoelia*, x 1; G, *Guadalupia*, x 1. **Bryozoa:** H, *Acanthocladia*, x 3. **Brachiopods:** I, *Eliva*, top and side views, x 1; J, *Strigirhynchia*, x 1.5; K, L, *Fascicosta*, top and front views, x 1.5; M, *Aphaurosia*, top and front views, x 1; N, *Paraspiriferina*, x 1; O, *Ariontha*, x 1; P, *Anteridocus*, top and front views, x 1.5; Q, *Dielasma*, front and top views, x 1; R, *Plectelasma*, front and top views, x 1; S, *Dyoros*, x 1; T, V, *Meekella*, end and top views, x .75; U, *Liosotella*, bottom and top views, x .75.

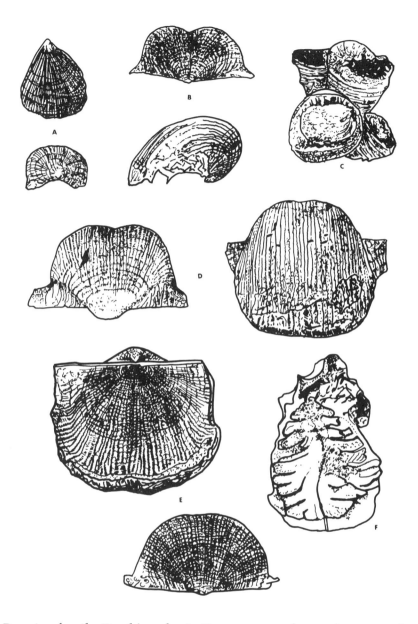

Permian fossils. **Brachiopods:** A, *Compressoproductus*, bottom and end views, x 1; B, *Thamnosia*, end and side views, x 1; C, *Prorichtofenia*, a cluster of four specimens, x 1; D, *Peniculauris*, end and bottom views (the slit-shaped holes in the bottom shell are the borings of barnacles), x 1; E, *Reticulatia*, top and end views, x 1; F, *Collemataria* (= *Leptodus*), view of interior of bottom valve, x 1; this aberrant brachiopod had a much reduced upper valve and an expanded lower valve that was open to the environment.

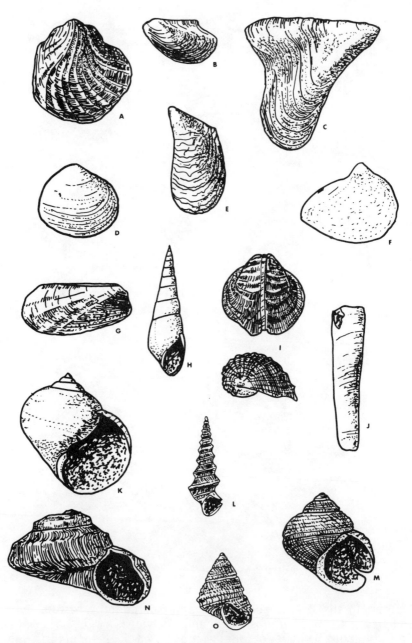

Permian fossils. **Pelecypods:** A, *Pseudomonotis*, x 1; B, *Dozierella*, x 2; C, *Myalina*, x .5; D, *Manzanella*, x 2; E, *Septimyalina*, x .75; F, *Schizodus*, x 1; G, *Permophorus*, x 2. **Gastropods:** H, *Meekospira*, x 1.5; I, *Retispira*, top and side views, x 1; K, *Ianthinopsis*, x 1; L, *Goniasma*, x 1.5; M, *Shansiella*, x 1; N, *Omphalotrochus*, x 1; O, *Tapinotomaria*, x 2. **Scaphopod:** J, *Plagioglypta*, x 1.

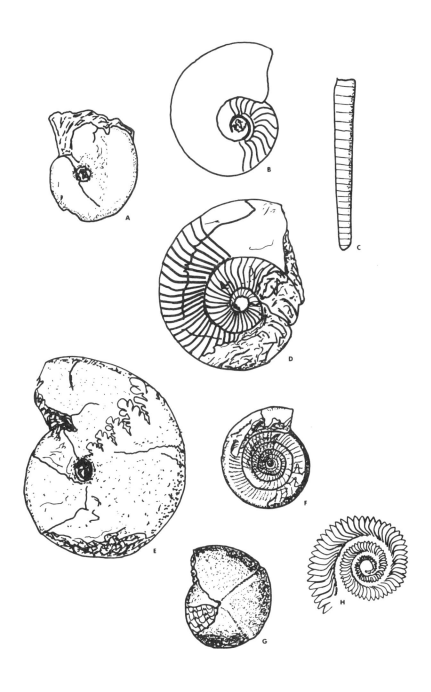

Permian fossils. **Nautiloids:** A, *Coelogastrioceras*, x .5; B, *Domatoceras*, x .3; C, *Mooreoceras*, x 1.5; D, *Stearoceras*, x .5. **Ammonoids:** E, *Perrinites*, x 1; F, *Metalegoceras*, x .75; G, *Peritrochia*, x .75. **Shark teeth:** H, *Helicoprion*, x .4.

THE NEW MEXICO FOSSIL RECORD

Triassic

Marine life during the Triassic Period (195–230 million years ago) reflects the reduced diversity occurring in the aftermath of the Permian extinctions. The main benthonic groups were pelecypods and gastropods, which, because of greater ecological tolerances were not as hard hit as other groups by the Permian crisis. Ammonoids recovered from near-extinction to evolve rapidly into hundreds of species, reaching a higher level of success than at any time during the Paleozoic, but brachiopods, bryozoans, crinoids, and corals were sparse. In North America marine environments were limited to the western United States and Canada; tectonic activity that had begun in the late Paleozoic erected an island arc system in this sea that stretched from Alaska to California. New Mexico and surrounding areas were the site of an immense alluvial plain, through which rivers flowed, spreading sediment and gradually transporting it to the western sea.

No fossiliferous marine Triassic rocks are preserved in New Mexico, and the state's fossil record for this period is limited to Late Triassic plants, vertebrates, and a few freshwater molluscs. The Triassic was a pivotal time in the evolution of amphibians and reptiles. For the amphibians, this was the last gasp of the large primitive forms; they became extinct at the end of the period. Many localities in New Mexico have yielded the corrugated polygonal scutes that protected the top of their heads; complete skulls and skeletons are occasionally found. The most impressive deposit was an "amphibian graveyard" about 25 km south of Lamy excavated in the 1930s. Here, a restricted bed containing numerous densely concentrated skeletons of *Metoposaurus*, a 2-meter-long genus, provided mute testimony to a pond that dried up, leaving these water-dwelling amphibians to die.

Many kinds of reptiles also inhabited New Mexico during the Late Triassic. Large quadrupedal thecodonts, an early group of archosaurs, had replaced the more primitive Permian reptiles. Some of these, like *Typothorax*, were covered with large bony armor plates. Others, like the phytosaurs, resembled crocodiles (though there is no close relationship) and they were the giants of the Triassic, reaching lengths of 7 meters and more. Phytosaur remains are probably the most common New Mexico Triassic vertebrate fossils; bone fragments, large conical teeth, and corrugated bony scutes are the most frequently preserved skeletal parts. Phytosaurs lived in ponds and streams very much as modern crocodiles do, feeding on amphibians and fish. Fish fossils are

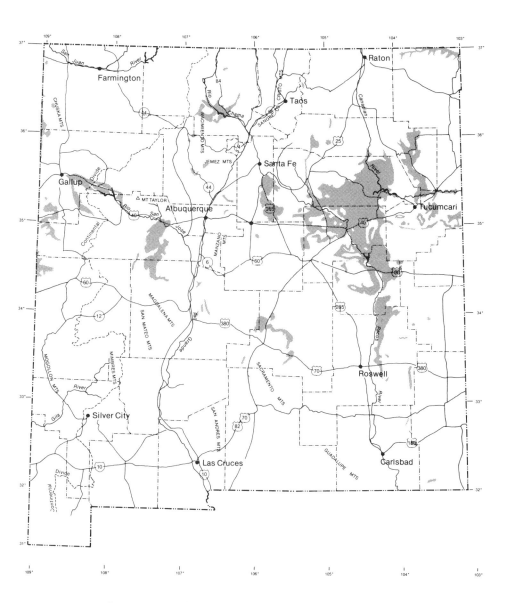

Outcrops of Triassic rocks in New Mexico.

THE NEW MEXICO FOSSIL RECORD

Skulls of *Metoposaurus* (called *Eupelor* or *Buettneria* in some publications) excavated from the Late Triassic "amphibian graveyard" south of Lamy. Each skull measures about fifty cm or more in length. *(From Colbert & Imbrie, 1956)*

sparse in the New Mexico Triassic; most of them consist of the scales and teeth of lungfish and coelacanths. In addition to body fossils of vertebrates, several different types of reptilian footprints have been reported from the Cimarron River area in Union County, not far from the New Mexico–Oklahoma border.

The Triassic was the time when both mammals and dinosaurs appeared—the two groups that were to prove to be the most successful of all terrestrial vertebrates. New Mexico Triassic rocks have not produced any remains of the earliest small mammals, but one of the best deposits in the world of early dinosaurs was discovered in 1947 near Ghost Ranch by an American Museum of Natural History expedition. From this locality were extracted numerous complete skeletons of *Coelophysis,* an agile, bipedal, carnivorous little dinosaur about 1.5 meters high, which was recently selected to be New Mexico's official state fossil. Curious-

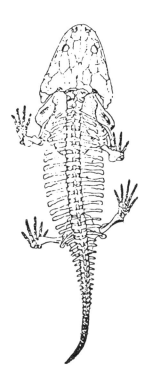

Skeletal reconstruction of *Metoposaurus*. This amphibian has been found in many Triassic localities in New Mexico and was apparently quite abundant around bodies of fresh water in north-central New Mexico about 200 million years ago.

ly, *Coelophysis* had first been described by Cope from a few fragments found in the Ghost Ranch area in the 1880s by David Baldwin. Study of the Ghost Ranch *Coelophysis* skeletons has contributed greatly to our understanding of the early history of dinosaurs, and illustrates the principle that, although the fossil record is incomplete, new discoveries continually add to our knowledge of the history of life, and to our knowledge of the evolution of particular groups. Besides Ghost Ranch, Triassic vertebrate fossils have been found in Cobre Canyon, near Ft. Wingate, near Gallina, south of Lamy, near Santa Rosa, around Mesa Redonda and Apache Canyon in Quay County, south of San Jon, and at several other localities in the Tucumcari area.

Other Triassic fossils found in New Mexico include plants and occasional freshwater pelecypods. The pelecypods are unionids, a group that persists today in streams and ponds. Plants of Triassic

THE NEW MEXICO FOSSIL RECORD

Reconstruction of a Late Triassic environment in the southwestern United States. At lower left is *Metoposaurus*; upper left, *Coelophysis*; lower right, *Rutiodon*, a large phytosaur about eight meters long. The plants along the pond are sphenophytes; the conifers are *Araucarioxylon*. (From an illustration by Margaret Colbert; reproduced with the permission of the W. W. Norton Co.)

age were first discovered in 1859 by J. S. Newberry during an early exploring expedition through northern New Mexico. Since then, especially within the last twenty years, good plant assemblages have been described from near Ft. Wingate, El Cobre Canyon, and several places in the Santa Rosa–Tucumcari area. The main types are ferns, conifers (particularly the most common Triassic petrified wood, *Araucarioxylon*, from a tree similar to the modern Norfolk pine), cycad leaves (like *Zamites*), and other gymnosperms. The presence of these plants, as well as amphibians and the water-dwelling phytosaurs in New Mexico Late Triassic deposits such as the Chinle Formation and Dockum Group, suggests that the prevailing climate was warm and moist. The nearest modern relatives of the plants live now in tropical or subtropical forests. Fossiliferous Triassic exposures continue from New Mexico into northeastern Arizona, where great numbers of highly col-

Triassic fossils. **Pelecypod:** A, *Unio*, a freshwater pelecypod, x 1. **Plants:** B, *Clathropteris*, a fern, x .5; C, *Zamites*, a cycad, x .5; D, *Otozamites*, a cycad, x .15; E, *Pagiophyllum*, a conifer, x 3; F, *Cynepteris*, a fern, x 2.; G, *Dinophyton*, a probable conifer, x 2.

THE NEW MEXICO FOSSIL RECORD

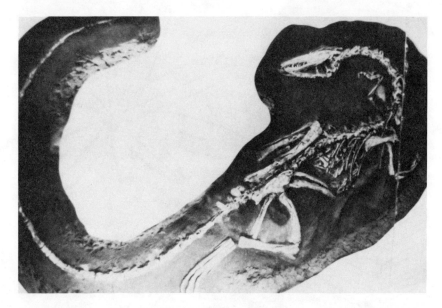

A cast of one of the *Coelophysis* skeletons excavated from Ghost Ranch in the 1940s. Such exceptionally well preserved vertebrate skeletons are rare in the fossil record, but provide an enormous amount of information about the structure and probable living habits of ancient vertebrates.

ored petrified tree trunks (mostly *Araucarioxylon*) as well as a variety of vertebrates are present in and around Petrified Forest National Park.

Jurassic

Sedimentary rocks of Jurassic age (140–195 million years ago) are exposed only in northern New Mexico and are limited to sediments representing terrestrial and hypersaline lagoonal environments. We know from marine fossils in other parts of the world that in the Jurassic seas many new groups were rising to prominence, and the composition of marine life was changing considerably from that of the Triassic. Because there are no Jurassic marine fossils in New Mexico, we will not dwell at length on

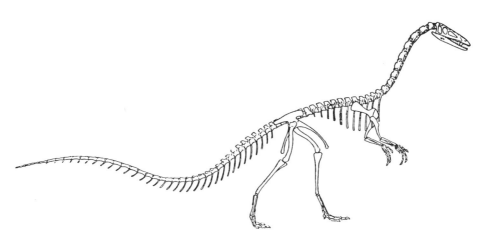

Skeletal reconstruction of *Coelophysis* in living position. The entire animal was between two and three meters long.

the nature of marine organisms during this period, but a few important points should be mentioned, since many of the evolutionary trends shown by Jurassic groups were to reach fruition in the Cretaceous, which is well represented in New Mexico.

Ammonoids, having nearly become extinct for the second time at the end of the Triassic, survived and became even more successful. Another group of cephalopods, the squidlike belemnites, became abundant for the first time during the Jurassic. Pelecypods and gastropods remained the most conspicuous inhabitants of the seafloor. The first small scleractinian coral reefs developed, and new groups of bryozoans, echinoids, and crinoids appeared that would consolidate their success in later times and continue to the present. The most advanced group of bony fish, the teleosts, evolved gradually from more primitive and more heavily armored forms. Simultaneously, essentially modern, fast-swimming sharks

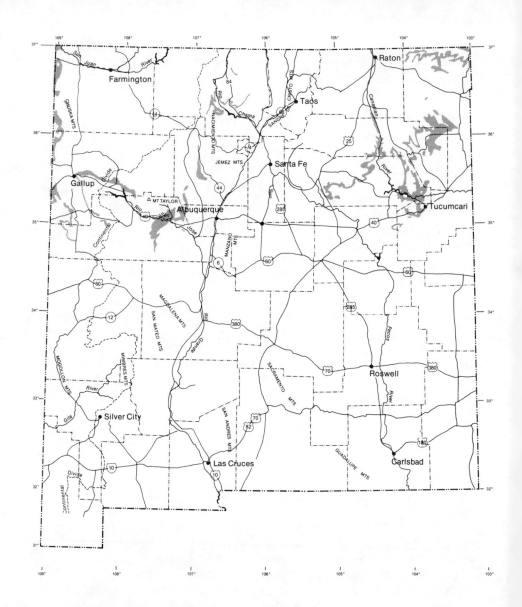

Outcrops of Jurassic rocks in New Mexico.

JURASSIC

A Jurassic fish, *Pholidophorus*, from the Todilto Formation. About actual size.

appeared, together with the rays, which were adapted to a bottom-dwelling existence. Several reptilian groups had become fully aquatic by the Jurassic, notably the plesiosaurs, ichthyosaurs, and some marine crocodilians. In short, the Jurassic was a time when the typical character of Mesozoic marine faunas was largely established; some of the new groups would vanish at the end of the era, but a good number continued on past the Cretaceous to remain important parts of modern marine life.

During the Jurassic, seas advanced and retreated several times over the western part of North America. A possible lagoon or enclosed extension of the sea existed in northern New Mexico and a sequence of gray and white limestones and gypsum, the Todilto Formation, was deposited in it. Fossils are very rare in the Todilto, consisting only of a few kinds of small fishes and one species of ostracod. The absence of all of the typical Jurassic marine invertebrates found farther west strongly suggests that conditions in the Todilto lagoon were inimical to most forms of sealife. The presence of gypsum, which precipitates out of very salty (hypersaline) water, leads to the conclusion that the water was too salty for most kinds of life. Just as the highly salty Great Salt Lake and Dead Sea are nearly devoid of aquatic organisms today, so was the Todilto Lagoon.

Later, near the end of the Jurassic, varicolored alluvial plain and river channel sediments were deposited over a large area in the central-western United States, including northern New Mexico. These terrestrial deposits, the Morrison Formation, have yielded some of the best Jurassic dinosaur fossils in the world, in Montana, Wyoming, Utah, and Colorado. Many of the skeletons prominently on display in large natural history museums in the eastern United States were extracted from the Morrison, and Dinosaur National Monument, in extreme eastern Utah, preserves intact over 1,000 bones on a tilted ancient stream channel. Many of the best dinosaur localities are local bone concentrations that developed as a result of carcasses piling up over months or years and quickly being buried on a sandbar or other sedimentological trap.

Many of the most familiar types of dinosaurs are found in the Morrison Formation, and these fossils indicate that dinosaurs had evolved considerably and had become much larger since their Triassic beginnings. Large herbivorous sauropods measuring 30 meters or more in length are common; such types as *Apatosaurus* (formerly *Brontosaurus*), *Camarasaurus*, and *Diplodocus* were the largest terrestrial animals that have ever lived. It has been traditionally thought that their mass (some weighed 30 or 40 tons) was so great that they could walk on land only with difficulty, and therefore spent most of their lives in the water, which helped to support their great bodies. Newer interpretations, however, suggest that they were not limited to an aqueous habitat but moved easily, though heavily, over the land, sometimes in herds. They must have spent their lives almost continuously eating to provide the energy to maintain their life processes, particularly the muscle power required for movement.

"Bird-hipped" dinosaurs are represented in the Jurassic by *Camptosaurus*, a relatively small herbivorous bipedal form, and by the bizarre *Stegosaurus*, which was quadrupedal (though a bipedal ancestry is indicated by the large size of the hind legs relative to the forelimbs) and possessed two rows of large bony plates along its back, with two pairs of long spikes near the end of the

JURASSIC

A life-sized model of *Stegosaurus*, guarding the entrance to Dinosaur National Monument in Utah. This dinosaur was about seven meters long. Isolated bones of *Stegosaurus* have been reported from the Morrison Formation near Mesa Gigante, between Albuquerque and Grants.

Reconstruction of *Antrodemus*, a 10-meter-long carnivorous dinosaur. A few teeth and bones of *Antrodemus* have been found in west-central New Mexico.

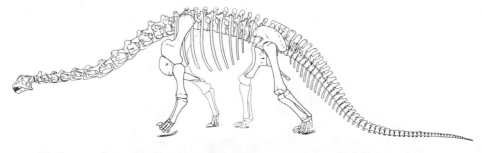

Skeleton of *Camarasaurus*, about 20 meters long. (After Marsh) About one-fourth of the skeleton was extracted from the San Ysidro site.

Dr. J. Keith Rigby, Jr., excavating a vertebra of the sauropod *Camarasaurus* near San Ysidro. Despite being petrified, fossil vertebrate bones are often brittle and fragment easily, so care must be taken in their excavation.

Camarasaurus vertebra after preparation in the laboratory. Parts of this bone have been reconstructed with plaster.

tail. Though the presumed function of this armor was protection for the backbone and spinal cord, it would seem that *Stegosaurus* would have been left open to flank attacks along the relatively unprotected sides of the body. Thus, some authorities have suggested that the plates did not stand erect on the back, but sloped downward over the sides. Others have argued that the primary function of the plates was not protective, but that they were used for dissipation of excess heat from the body. Both the sauropods and "bird-hipped" dinosaurs had small teeth that were probably best suited for eating soft vegetation.

Large carnivorous dinosaurs (carnosaurs) are also conspicuous in the Morrison Formation. The best known example is *Antrodemus* (formerly *Allosaurus*), a bipedal genus 10 or more meters long and 5 meters high, which fed primarily on the large herbivores. Much smaller, more agile types (coelurosaurs) resembling plucked ostriches probably preyed upon lizards, early mammals, the young of other dinosaurs, and possibly insects. A characteristic feature of carnivorous dinosaurs is the great reduction of the front limbs relative to the hind legs, and the development of sharp serrated teeth for slicing easily through their prey. Whether the large carnosaurs were active and efficient hunters or subsisted mainly as scavengers has been debated by paleontologists; probably they did some of both.

Along the forested streams that wound through the Jurassic landscape, crocodiles and turtles lived and even 150 million years ago these reptiles looked fairly similar to modern members of these groups. True birds, and the unrelated flying pterosaurs, also appeared in the Jurassic, though birds are not present and pterosaurs are rare in the Morrison.

In northern New Mexico the Morrison Formation is widely exposed and lithologically similar to the dinosaur-bearing beds farther north, but until recently only scattered small bone fragments and plants had been found. The plants, generally tree trunks and branches that are petrified or impregnated with uranium ore were too poorly preserved to be identifiable. In the early 1950s bone fragments of dinosaurs identified as *Stegosaurus*, *Brontosaurus* (now *Apatosaurus*), and *Allosaurus* (now *Antrodemus*) were discovered by a University of New Mexico geology student in the Grants-Acoma area west of Albuquerque. These finds suggested that the Morrison Formation might be richer in dinosaur remains than had previously been suspected. This possibility was given further support by recent discoveries north of Bernalillo. In the spring of 1979, an amateur paleontologist reported a couple of complete dinosaur bones in an arroyo just west of San Ysidro. Dr. Keith Rigby, vertebrate paleontologist for the Bureau of Land Management, authorized further excavation of the site, and with the help of a backhoe and numerous members of local rock-hound groups, about one-quarter of the skeleton of a 20-meter-long *Camarasaurus* was extracted, along with a few serrated teeth from a carnivorous dinosaur. Since then other bone concentrations have been located in the Grants area, and it seems likely that with further searching New Mexico may eventually yield its share of Jurassic dinosaurs. With nearly half of the country's uranium being mined from the New Mexico Morrison it is particularly important that dinosaurs exposed during mining operations be brought to the attention of paleontologists.

Cretaceous

During the Cretaceous Period (65–140 million years ago) the last great marine transgressions covered at least one-third of the world's land area; the earth would never again be such an oceanic world. The enlarged oceans soaked up heat from the sun and their currents distributed it to all latitudes, resulting in an overall warm climate from the equator to the poles. No ice caps existed in the Cretaceous; dinosaurs lived north of the Arctic Circle and magnolia trees grew in Greenland. In North America, by about 100 million years ago, a great shallow sea covered the central part of the continent from the Gulf of Mexico to Canada, and part of its western shoreline ran right through New Mexico. The inhabitants of this sea are abundantly preserved in New Mexico, and in the later Cretaceous the remains of extensive forests and a wide variety of terrestrial animals also come into the record as the shoreline shifted back and forth across the northern and central parts of the state.

Marine fossils of Early Cretaceous age are present in southern and eastern New Mexico, and were first observed in 1853 by the Swiss geologist Jules Marcou around Pyramid Mountain, near the present town of Tucumcari. The first new species ever named from New Mexico (an oyster named *Texigryphaea tucumcarii*) came from this area. In addition, fossiliferous Early Cretaceous beds many thousands of feet thick are known in the southwestern panhandle (Big and Little Hatchet Mountains), and somewhat thinner sequences are exposed around Cerro de Cristo Rey, a mountain that straddles the border between New Mexico and

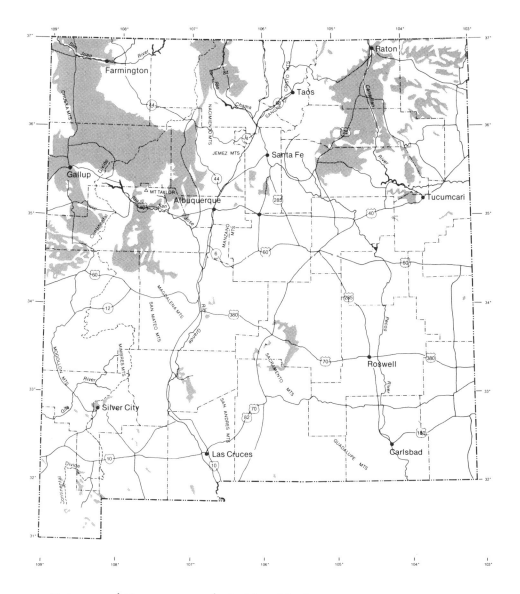

Outcrops of Cretaceous rocks in New Mexico.

Mexico just west of El Paso, and in several other places in southern New Mexico. The fossils in these rocks are predominantly ammonoids, pelecypods, and gastropods, testifying to the abundance and great success of molluscs in Cretaceous seas. The ammonoids possessed highly convoluted, extremely complex sutures, in contrast to the generally simple sutures seen in Paleozoic ammonoids. The pelecypods and gastropods had increased considerably in variety and abundance since we last saw them in the Permian; most of them are very modern-looking and, indeed, the origin of many modern families of these molluscs dates back to the Cretaceous. Particularly conspicuous among the pelecypods are several kinds of oysters (*Ostrea, Lopha, Texigryphaea*), pectens (e.g., *Neithea*) and a bizarre group that had a short but spectacular career, the rudistids.

Rudistids, like some late Paleozoic brachiopods, evolved a conical shape, with one shell becoming very large (some are a meter high) and the other being reduced to a small cap. The shells of these large pelecypods are very porous and easily recognizable even as fragments in Early Cretaceous rocks. Rudistids are of interest because they often grew on the seafloor in high densities, forming reefs in some places that perhaps competitively challenged the conventional scleractinian coral-algal reefs. The rudistids, however, became extinct at the end of the Cretaceous, whereas coral reefs are still abundant.

Several other groups are less common in the Early Cretaceous of New Mexico. Echinoids are sometimes found exquisitely preserved, with the spherical test intact and each plate and pore visible, though without the easily detached spines. Both regular and flattened "irregular" echinoids are present. Scaphopods such as *Dentalium* have been collected in the Big Hatchet Mountains; they look fairly similar to modern members of this genus. *Orbitolina* is the most common of several genera of foraminifers; it is disc-shaped to broadly conical and measures about one-half centimeter in diameter. Large numbers of these forams occur in some Early Cretaceous limestones in the Big Hatchet Mountains. A few corals are also known from New Mexico Early and Late Cretaceous deposits, but apparently the Midcontinent sea was not favorable to their growth, for occurrences are scattered and sparse. Large Cretaceous coral reefs are found elsewhere, for example, in Texas, Mexico, Saudi Arabia, and the Mediterranean area, and many of them are filled with oil. Calcareous algae, another important constituent of reefs, are locally abundant, but rarely well preserved, in some New Mexico Early Cretaceous beds. Finally,

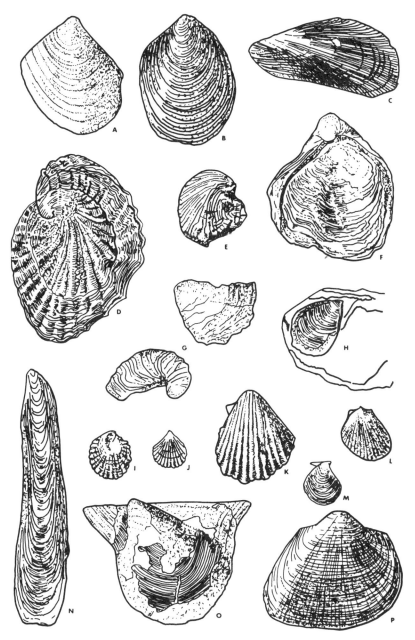

Cretaceous fossils. **Pelecypods:** A, *Inoceramus*, x 1; B, *Mytiloides*, x 1; C, *Modiolus*, x 1; D, *Ceratostreon*, x .75; E, *Exogyra*, x .75; F, *Texigryphaea*, x .5; G, *Pycnodonte*, bottom and side views, x .75; H, *Pseudoperna*, x 1; I, *Lopha*, x 1; J, *Plicatula*, x 1; K, *Neithea*, x 1; L, *Lima*, x 1; M, *Camptonectes*, x 1; N, *Crassostrea*, x .25; O, *Phelopteria*, x .75; P, *Idonearca*, x 1.

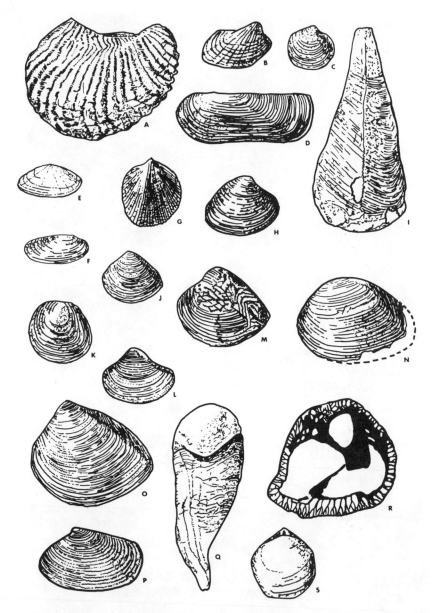

Cretaceous fossils. **Pelecypods:** A, *Scabrotrigonia*, x 1; B, *Pholadomya*, x 1; C, *Lucina*, x 1; D, *Leptosolen*, x 1; E, *Tellina*, x 1; F, *Legumen*, x 1; G, *Granocardium*, x 1; H, *Aphrodina*, x 1; I, *Pinna*, x .5; J, *Cymbophora*, x 1; K, *Anomia*, x 1; L, *Psilomya*, x 1; M, *Quadrula*, a freshwater genus, x 1; N, *"Unio,"* a freshwater genus, x 1; O, *Trigonarca*, x 1; P, *Anatina*, x 1; Q, Caprinid Rudistid, x .25; R, Caprinid Rudistid, cross-sectional view, x .75. **Brachiopod:** S, *Kingena*, x 1.

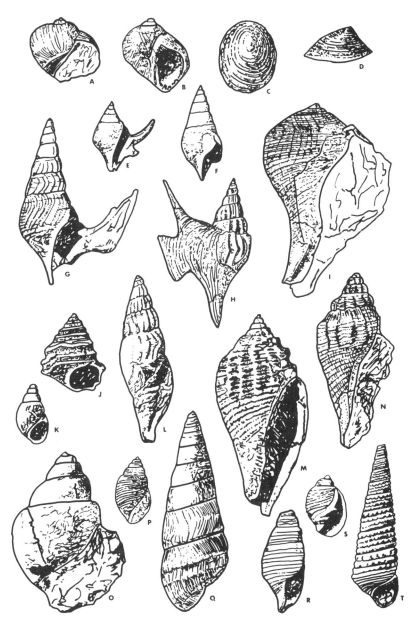

Cretaceous fossils. **Gastropods:** A, *Gyrodes*, x 1; B, *Lunatia*, x 1; C, D, *Anisomyon*, top and side views, x 1; E, *Lispodesthes*, x 1; F, *Arrhoges*, x 1; G, *Anchura*, x 1; H, *Perissoptera*, x 1; I, *Pyropsis*, x 1; J, *Tulotoma*, a freshwater genus, x 1; K, *Campeloma*, a freshwater genus, x 1; L, *Paleopsephaea*, x 1; M, *Volutomorpha*, x 1; N, *Carota*, x 1; O, *Tylostoma*, x .5; P, *Actaeon*, x 1.5; Q, *Cassiope*, x 1; R, *Vascellum*, x 1.5; S, *Ringicula*, x 3; T, *Turritella*, x 1.

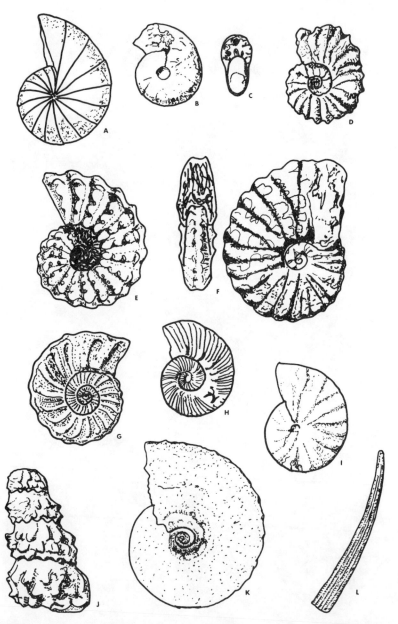

Cretaceous fossils. **Nautiloid:** A, *Eutrephoceras*, x .25. **Ammonoids:** B, C, *Desmoceras (Pseudouhligella)*, side and front views, x 1; D, *Tarrantoceras*, x 1; E, *Mortoniceras*, x 1; F, *Metoicoceras*, front and side views, x .75; G, *Collignoniceras*, a small specimen, x 1; H, *Prionocyclus*, a small specimen, x 1; I, *Coilopoceras*, a small specimen, x 1; J, *Turrilites*, x 1; K, *Placenticeras*, x 1. **Scaphopod:** L, *Dentalium*, x 1.

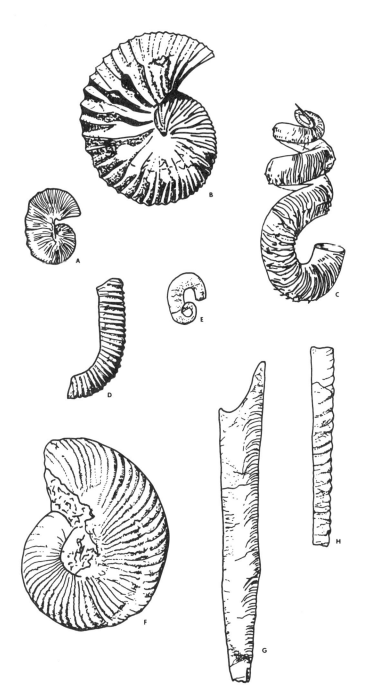

Cretaceous fossils. **Ammonoids:** A, *Scaphites*, x .75; B, *Clioscaphites*, x 1; C, *Didymoceras*, x .25; D, *Glyptoxoceras*, x 1; E, *Worthoceras*, x 1; F, *Oxytropidoceras*, x .25; G, *Baculites*, x 1; H, *Sciponoceras*, x 1.

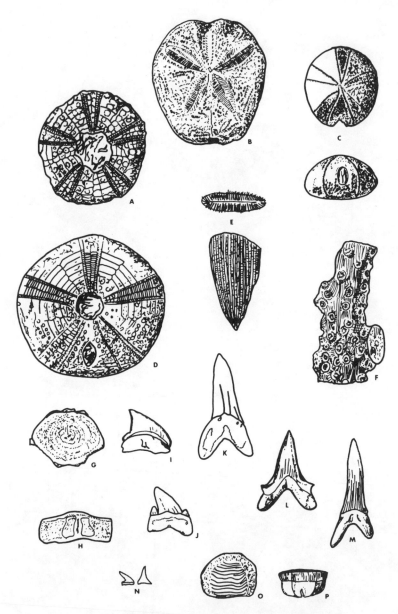

Cretaceous fossils. **Echinoids:** A, *Phymosoma*, bottom view, x 1; B, *Hemiaster*, top view, x 1; C, *Conulus*, top and end views, x 1; D, *Coenholectypus*, bottom view, x 1. **Scleractinian Corals:** E, *Aulosmilia*, top and side views, x 1; F, *Archohelia*, x .5. **Bony Fish:** G, H, *Amia* vertebra, front and top views, x 1. **Shark teeth:** I, *Squalicorax*, x 1; J, *Otodus*, x 1; K, *Isurus*, x 1; L, *Lamna*, x 1; M, *Scapanorhynchus*, x 1; N, *Squatina*, x 1. **Ray teeth:** O, *Ptychodus*, x .5; P, *Myledaphus*, usually found in nonmarine sediments, x 1.

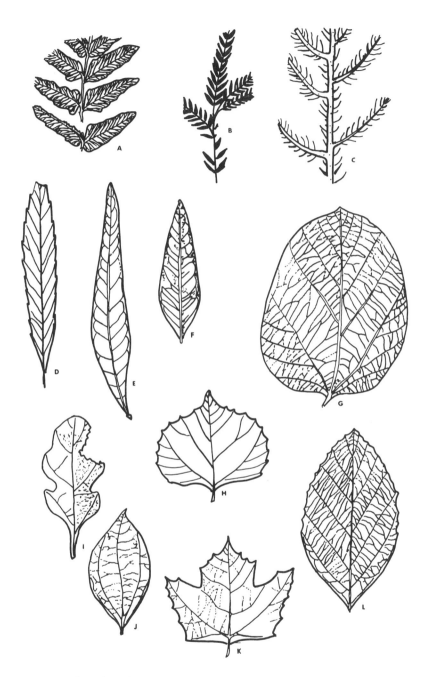

Cretaceous fossils. **Plants** (are all × 1; all but A–C are angiosperms): A, *Osmunda*, a fern; B, *Sequoia*, a conifer; C, *Araucaria*, a conifer; D, *Salix*; E, *Dryophyllum*; F, *Sapindus*; G, *Ficus*; H, *Populus*; I, *Quercus*; J, *Cinnamomum*; K, *Platanus*; L, *Viburnum*.

THE NEW MEXICO FOSSIL RECORD

Reconstruction of a Late Cretaceous seafloor, showing large coiled ammonoid *(Placenticeras)* at left; numerous straight conical ammonoids *(Baculites)*; a small scaphitoid ammonoid, center foreground; an uncoiled ammonoid *(Didymoceras)*, center; and numerous types of pelecypods and gastropods (see line drawings for names). *(Courtesy of the Smithsonian Institution)*

Early Cretaceous deposits in southern New Mexico have yielded good remains of the enigmatic fern *Tempskya*, which had a "false trunk" 30 cm or more in diameter with typical fern foliage.

Late Cretaceous sedimentary rocks are exposed in most parts of the state; the most extensive outcrops are in the San Juan and Raton basins, in northwestern and northeastern New Mexico respectively, but important fossiliferous beds are also present on the east side of the Sandia Mountains, around Galisteo, northwest and southeast of Socorro, and along the Rio Puerco, a few miles west of Albuquerque. In general, the main fossil groups are the same as in the Early Cretaceous, but the faunas differ in their specific compositions. Ammonoids—dozens of species—are abundant and highly varied. Relatively small, partially uncoiled forms *(Scaphites)* are among the most common, but giants up to a meter in diameter (e.g., *Coilopoceras, Placenticeras*) are present in moderate abundance at some localities. Many ammonoids developed pronounced knobs or spikes on their shells (e.g., *Acanthoceras*), while others are almost completely uncoiled (like *Baculites*) and resemble superficially the long extinct straight-coned nautiloids. Though extremely diverse and successful through the Cretaceous, ammonoids became extinct at the end of the period, though strangely, a few types of coiled nautiloids survived and still live in modern oceans. The rise of advanced ray-finned fish (teleosts) and sharks in the Cretaceous, which may have outcompeted the ammonoids in their predaceous, swimming life-style, possibly contributed to the downfall of this long-lived group.

Pelecypods are ubiquitous in the Late Cretaceous. Oysters such as *Ostrea, Pycnodonte, Lopha,* and *Exogyra* are extremely common and can be found by the thousands in some beds. *Pinna*, with its unusual triangular shape, is sometimes found in the center of rounded dark concretions that erode out of some Late Cretaceous shales (ammonoids are less common "nuclei" in such concretions). *Pinna,* and genera like *Pholadomya, Idonearca, Tellina,* and *Granocardium* closely resemble analogous modern pelecypods.

Along with oysters, the most characteristic pelecypods are the inoceramids (e.g., *Inoceramus,*), which were amazingly successful during the Cretaceous, but became extinct at the end of the period. Inoceramids have a roughly oval shape and strong concentric ridges on the shell. About two dozen species, which are distinguished by minor differences in shape, size, and ornamentation pattern, are known from New Mexico, and well over 100 species have been described from the area of the Midcontinent sea. The largest New Mexico specimens reach perhaps 30 cm in diameter, but some types found in Kansas have shells several times this size that resemble, according to one paleontologist, shallow bathtubs. Small fragments of broken inoceramid shells are very distinctive though sometimes perplexing constituents of some marine sediments; they appear as chips of elongate, almost fibrous calcite elements up to a centimeter thick.

Perhaps the strangest Late Cretaceous pelecypods are "shipworms" that bored through branches and trunks of trees (modern forms prefer ships and pier pilings), producing sinuous passageways up to about a centimeter wide that are occasionally seen in petrified wood, and nearshore marine and brackish sediments.

Most gastropods observed in the Late Cretaceous of New Mexico are also similar to modern forms; a concentration of snails in a Cretaceous shoreline sandstone will be immediately recognized

Coilopoceras, one of the largest ammonoids found in New Mexico. This specimen, from the Rio Puerco area west of Albuquerque, is about 75 cm in diameter.

A beautifully preserved specimen of *Acanthoceras,* an ammonoid with large knobs projecting from the shell. The complex sutures may clearly be seen on the earlier whorls. Original is about 20 cm wide. From the Upper Cretaceous of northwestern New Mexico.

The trace fossil *Ophiomorpha,* the burrow of a small shrimp. Note the knobby texture of the burrow. *Ophiomorpha* is a common and distinctive fossil in the Dakota Sandstone and other sandstone formations in northern New Mexico.

Skeleton of a mosasaur (from Williston). Original about 10 meters long. A partial skeleton of a similar form was recently discovered in northern New Mexico.

as such by anyone who has collected shells along modern beaches. Worthy of note here are the thin, high-spired *Turritella,* and the large, elegantly ornamented *Volutomorpha;* both are common in New Mexico.

Many other animals shared the Late Cretaceous sea with the dominant molluscs. Microscopic foraminifers are abundant in some formations, and these include for the first time planktonic forms as well as bottom-dwelling types. Ovoid smooth brachiopods are rare, and bryozoans have been reported only a couple of times, encrusting skeletal fragments of other organisms. Large crustaceans are likewise rare; one crab carapace and a few pieces of shrimps and lobsters are all that have ever been found, but ostracods are moderately common locally. Indirect remains of shrimp, however, are common in some sandstones that represent beach or very nearshore environments. These fossils consist of filled-in bifurcating burrows with a knobby external surface—the trace fossil *Ophiomorpha.* As noted previously, corals are very uncommon, though a thicket of them grew on the seafloor near Lamy.

Well-preserved marine vertebrates are rare in the New Mexico Late Cretaceous. We know from many complete skeletons in Kansas, South Dakota, and adjacent states, that advanced ray-finned fish, some of them five or more meters long, were common in the Midcontinent sea, but only a few teeth and scales are known from New Mexico deposits. Shark and ray teeth, however, are sometimes abundant, mainly in sandstones. Several kinds of sharp, single- or multicusped shark teeth indicate the presence of advanced sharks nearly identical to predaceous modern types. Ray teeth are relatively large, with a blunt massive cusp that aided

A palm frond impression *(Sabalites)*. Original is about 40 cm long. These are moderately common in some Late Cretaceous formations; petrified palm stumps and roots, distinguished by the presence of many small vascular bundles, are also occasionally found.

the rays in crushing the invertebrates that were the major portion of their diet.

Aquatic reptiles—plesiosaurs and mosasaurs—were well represented in the Midcontinent sea, though few remains have been found in New Mexico. Plesiosaur bones were first collected from northwestern New Mexico nearly 100 years ago; in the 1960s a row of several vertebrae was reported from the Burro Mountains southwest of Silver City. Similarly, mosasaurs, marine lizards up to 15 meters long, were known from a few bones and teeth from the Jornada del Muerto area and near Raton, but in 1979 a partial skeleton consisting of more than forty vertebrae, several ribs, and jaw and tooth fragments, was collected from marine shales southeast of Dulce. These recent discoveries are promising and it is likely that eventually some lucky collector will find a fairly complete skeleton of one of these magnificent reptiles.

CRETACEOUS

A petrified log in the Bisti badlands. Such logs are common in some areas of the Fruitland-Kirtland badlands; some are up to 30 meters long. Most of the logs are from conifers.

Perhaps the rarest New Mexico Cretaceous fossils are those of birds. From well-preserved skeletons in Kansas we know that swimming and flying birds, some fairly large and having teeth set in their beaks, were relatively common in some places on the eastern side of the Midcontinent sea. The only Cretaceous bird fossil ever found in New Mexico, however, is a hollow piece of limb bone about three centimeters long that was discovered in marine sediments near Cuba about two years ago.

During the Late Cretaceous, the New Mexico shoreline of the Midcontinent sea shifted back and forth in a northeast-southwest direction several times before the sea finally left the state near the very end of the period. One result of these fluctuations was the deposition in the San Juan Basin of sediments representing many different environments, from fairly deep marine shales (like the Mancos Shale) to nearshore and beach sandstones to terres-

THE NEW MEXICO FOSSIL RECORD

Part of the "fossil forest," east of the Bisti Badlands in northwestern New Mexico. Several in situ stumps are visible; many dozens of them plus several large logs are distributed within a few acres at this site. The forest was discovered by the author and Keith Rigby during a paleontological survey of the San Juan Basin in 1977. It is in an area that will probably undergo strip mining in the near future.

trial swamp and river sediments. Exposures of the last environment, eroded into a desolate badlands topography, stretch in a great arc southward from near Farmington, then turn eastward just above Chaco Canyon, and extend all the way to Cuba.

Study of the fossils in these sediments, especially the Fruitland and Kirtland Formations, allows us to visualize in great detail the terrestrial life of northwestern New Mexico 65 to 70 million years ago. Extensive forests grew there, with ferns and conifers as well as a large variety of deciduous trees with leaves much like those of modern palm, magnolia, willow, fig, laurel, sycamore, and other trees. Deciduous trees and shrubs are flowering plants, and the Late Cretaceous marks the first entry of this now ubiquitous group into New Mexico's fossil record. The Fruitland-Kirtland flora indicates a warm, moist, well-watered environment, and it has a distinctly modern look. These plants are represented in the Bisti and neighboring areas by excellently preserved leaves and many petrified logs, some more than 30 meters in length. In one place, discovered in 1977 and informally named the "fossil forest," dozens of petrified stumps are preserved in their living positions, the remnants of a buried forest. Adding to the importance of this unusual occurrence is the fact that scattered among the stumps and in surrounding areas are the fossils of the animals that inhabited the forest. Much of the vegetation preserved in the Fruitland and Kirtland Formations has been transformed into coal, which has made these formations of interest to mining companies as well as to paleontologists. Excellent deposits of fossil plants also occur in Late Cretaceous exposures in the Raton–Vermejo Park area. Over 100 species have been re-

A complete turtle shell. Only the light area near the center was exposed when it was found, which is why the shell was intact, rather than fragmented by erosion. Several dozen complete shells have been collected from the Fruitland-Kirtland badlands over the years. The shell is about 60 cm long.

ported; in general they are similar to the plants of northwestern New Mexico, but strangely, vertebrate fossils are almost entirely absent.

Examination of other fossils in the Fruitland-Kirtland badlands adds to our picture of the Late Cretaceous environment in the San Juan Basin. In and near streams large numbers of turtles lived; shell fragments and occasionally complete shells are perhaps the most common vertebrate fossils in these badlands. Heavily pitted and corrugated crocodile scutes and conical teeth are also common. Shiny brown-purple diamond-shaped scales indicate large populations of garfish, closely related to modern species still living in ponds and streams of the southeastern United States. In fact, the presence of crocodilians, turtles, and garfish in the New Mexico Cretaceous brings to mind the Florida Everglades, where

Left: Scales from the garfish *Atractosteus* (= *Lepisosteus*). Concentrations of scales in channel sandstones in the Fruitland and Kirtland formations (as well as in Paleocene and Eocene formations) often indicate productive sites for other, smaller vertebrate remains, such as mammal teeth. The scales shown here are about natural size. Right: Crocodilian scutes are common fossils in Late Cretaceous, Paleocene, and Eocene deposits in the San Juan Basin. A single crocodile had hundreds of these bony plates underneath the skin along its back and sides. About natural size. Top: A crocodilian tooth. These relatively large conical teeth are fairly common in some areas of the San Juan Basin, and, with the numerous scutes, attest to the abundance of crocodilians in northwestern New Mexico 50 to 70 million years ago.

these same three groups, little changed over the past 70 million years, are still prevalent. A few teeth and bones of small lizards and snakes have also been reported from these deposits, but they were apparently not very common. Freshwater molluscs, including both unionid pelecypods and several species of gastropods, lived on the stream bottoms; the shiny mother-of-pearl layers of their shells sparkle in the sunlight as they erode out of the somber badlands.

The largest Fruitland-Kirtland vertebrates are the dinosaurs. In the badlands, fragments of their petrified bones are quite common, but hundreds of complete bones, and even some skulls and partial to complete skeletons, have been found over the past seventy-five years. The dominant Late Cretaceous dinosaurs in New Mexico and other localities in the northern Rocky Mountains region are considerably different from the main Late Jurassic

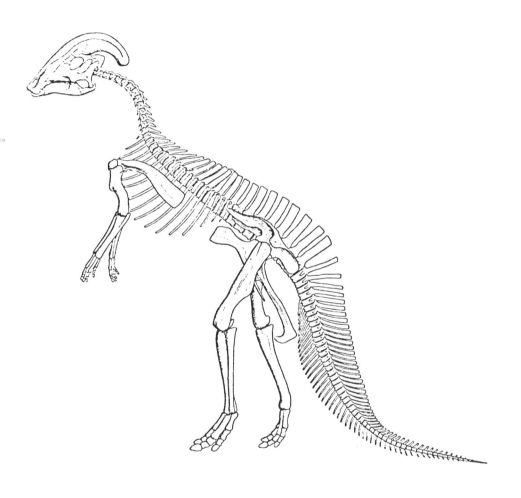

Skeletal restoration of *Parasaurolophus,* a crested hadrosaur about 15 meters long. This almost complete skeleton was excavated near the Bisti Badlands in the 1920s. *(From Ostrom)*

Skeleton and restoration of *Pentaceratops*, a five-horned dinosaur known only from New Mexico. The human figure has been added for scale. *(Courtesy of the Museum of Northern Arizona)*

types from the Morrison Formation. The giant sauropods, so common in the Jurassic, are rare in the Cretaceous; the first and only known Late Cretaceous North American sauropod, *Alamosaurus*, was first discovered near Ojo Alamo Spring east of Bisti. The dominant Fruitland-Kirtland dinosaurs are hadrosaurs (the so-called duck-billed dinosaurs) and ceratopsians (horned dinosaurs). Hadrosaurs were medium-sized bipedal herbivores about 10 meters long that were partially aquatic. Their jaws were equipped with a formidable battery of hundreds of teeth that collectively formed a flat surface for grinding up the pine needles and other vegetation used as food. One genus, *Parasaurolophus*, known from two good skeletons and numerous bones from New Mexico. possessed a curious hollow elongate crest extending from the back of the head.

The horned dinosaurs were massive five-to-eight-meter-long

CRETACEOUS

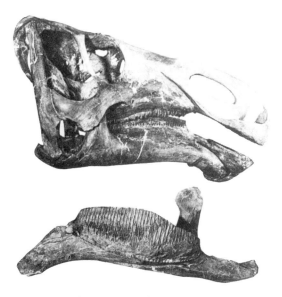

Skull and lower jaw of *Kritosaurus*. Several partial skeletons, such as the one excavated in 1979, have come to light in the last few years in the Late Cretaceous of the San Juan Basin. The skull illustrated, found in the early 1900s, shows only a slightly raised nasal area; we now know that the nasal area of this genus is more elevated than shown in this early restoration. Note the large number of teeth in the jaw. Isolated hadrosaur teeth, from *Kritosaurus* and *Parasaurolophus*, are relatively common finds in the Fruitland and Kirtland Formations. *(From Lull & Wright)*

herbivorous reptiles with large heads, a distinctive curved frill extending over the neck region and several horns projecting from the head. New Mexico's best-known horned dinosaur, *Pentaceratops*, had five horns, and is known from several two-meter-long skulls, a couple of nearly complete skeletons and numerous isolated bones. It has not been found anywhere outside of the state. Fragmentary remains indicate the presence of at least one other type of ceratopsian in the Fruitland and Kirtland Formations. A fourth group of New Mexico Late Cretaceous dinosaurs, the ankylosaurs, is known only from rare fragments of the heavy dorsal armor. Individual bones of the armor are about twenty cm long, oval in shape, and rough to the touch.

The dinosaurs discussed above were all plant eaters, but carnivorous forms roamed northwestern New Mexico as well. Both large carnosaurs related to *Tyrannosaurus* and small coelurosaurs

THE NEW MEXICO FOSSIL RECORD

A ceratopsian limb bone. Isolated bones like this are far more common than articulated skeletons. When exposed to the forces of erosion, however, even large fossil bones fragment rapidly, as this one is beginning to do.

were present, but remains are limited to their triangular serrated teeth, claws, and occasional bones. The large carnosaurs had their front limbs reduced to extremely small and probably nearly useless appendages. The low number of fossils of carnivorous dinosaurs, relative to herbivores, is not surprising—in most communities of organisms plant-eating animals are far more abundant than meat-eaters; many herbivores are required to keep each carnivore well fed over a given period of time.

Almost all Late Cretaceous dinosaur fossils in New Mexico have come from the San Juan Basin, but a few have been found elsewhere, though most of these are limited to unidentifiable fragments that had been washed into marine sediments. A ceratopsian skull was eroding from the McRae Formation many years ago when it was covered by the rising waters of Elephant Butte Reservoir before it could be collected.

CRETACEOUS

A single bony plate from the back of an ankylosaur, or armored dinosaur, found in the Bisti Badlands. About one-third natural size.

The last major group of vertebrates represented in the Late Cretaceous badlands are mammals. Advanced mammals had developed and expanded in the Cretaceous, but were still very small, possibly nocturnal or secretive creatures that probably spent a good deal of time trying to stay away from the large variety of reptiles with which they coexisted. Increasing numbers of almost microscopic mammal teeth and minute jaw fragments, obtained by carefully screen-washing large volumes of ancient stream channel sediments, have turned up as paleontologists have recently concentrated their search for these small but important fossils. Though relatively rare and inconspicuous in the Late Cretaceous, mammals were to undergo a rapid expansion at the beginning of the Paleocene that made them the dominant terrestrial vertebrates and ended the Mesozoic "Age of Reptiles."

The Fruitland and Kirtland formations were deposited near the

THE NEW MEXICO FOSSIL RECORD

A carnosaur tooth, from a dinosaur very similar to *Tyrannosaurus rex.* Note the serrations on the edge. The tooth is laterally compressed and bladelike; the photo is a little smaller than natural size.

close of the Cretaceous Period, and the dinosaurs preserved in these beds represent some of the last members of this 150-million-year-old group. A crisis for both terrestrial dinosaurs and the large marine reptiles occurred at the end of the Cretaceous, about 65 million years ago. At that time these groups and the pterosaurs, along with invertebrates like ammonoids, many types of foraminifers and the inoceramid and rudistid pelecypods rather abruptly became extinct. The cause or causes of this episode of extinctions have been much discussed; one source listed more than forty different serious and fanciful explanations that have been advanced at one time or another. Probably several factors were involved, including increasing seasonality (variation between winter and summer temperatures), changes in environment and climate caused by the retreat of the large mid-continental seas, and gradual but increasingly profound changes in the structure of

CRETACEOUS

Skeleton of the giant carnosaur *Tyrannosaurus rex*, which measures about 15 meters long and 7 meters high. Though this dinosaur has not been found in New Mexico, close relatives that were slightly smaller *(Albertosaurus)* are known from the Late Cretaceous dinosaur beds of northwestern New Mexico. *(From Osborn)*

terrestrial and marine communities caused by the progressive introduction of new organisms into these communities. These may have contributed to a decline in dinosaurs and other animals, with the coup de grace being supplied by a short-term catastrophic event.

One catastrophe for which evidence has recently been reported is the impact of a large (several kilometers diameter) asteroid on the earth at the end of Cretaceous time. Such an impact would have thrown enormous amounts of dust into the atmosphere, perhaps enough to have prevented most sunlight from reaching the earth's surface for a few years. Land plants and oceanic phytoplankton would have died (plants require sunlight to photosynthesize their food), followed by animals depending on the plants for food, and ultimately by carnivores depending on plant-eating animals. Shutting off most of the sunlight for just a few years

Dinosaur skeletons are constantly eroding out of the Late Cretaceous badlands of northwestern New Mexico. Here is part of the vertebral column of *Kritosaurus*, a duck-billed dinosaur, eroding out of a sandstone bed east of the Bisti area. The exposed parts of the skeleton (beneath the figure) are fragmented, but sufficient to indicate that there might be better preserved remains embedded in the rock.

conceivably could have been enough to cause the extinction of many previously successful groups. When the asteroid dust had settled and sufficient amounts of sunlight were again reaching the earth, vegetation would be reestablished from dormant spores and seeds.

One problem in explaining the extinction of the dinosaurs has always been: why did dinosaurs become extinct and not other vertebrates, such as mammals or other reptiles like crocodiles, turtles, and lizards? The animals that survived were all relatively small; large size apparently became a fatal handicap at the end of the Cretaceous. If drastic short-term climatic changes accompanied an asteroid impact, as is likely, smaller animals could find places to hide that would have afforded them some protection, but large animals had to endure the changes. Moreover, small animals that required relatively little food or which could subsist

The *Kritosaurus* skeleton shown to the left, discovered in 1977, was excavated in 1979 by the Museum of Northern Arizona. This photo shows the site after the surface bones were collected and excavation back into the sandstone had commenced. Part of the rib cage was exposed (to left of plaster-covered bone), as well as limb bones that had not been visible on the surface. An upper arm bone (humerus) has been encased in plaster of Paris to keep it from crumbling when it is removed. This bone, nearly a meter long, and other skeletal elements suggest that this dinosaur is one of the largest of its kind known from the Bisti area.

for a time on high-energy food like nuts and seeds stood a much better chance of enduring a food crisis than a dinosaur, which needed continuous access to enormous amounts of food.

    Whatever the exact causes of the Late Cretaceous extinctions, they helped set the stage for the development of modern faunas. As the Cenozoic Era began, the major players in this most recent act of earth history were in place. There would be changes to be sure; the processes of evolution would continue to ensure that new species would develop and others would become extinct, that groups would flourish and decline as the environment and their interactions with other groups changed, but there would be no more dramatic crises on the scale of the Permian and Cretaceous extinctions. A visitor transported by his imagination back to the world of the early Cenozoic would find much that was familiar to him there.

THE NEW MEXICO FOSSIL RECORD

Paleocene

The Cenozoic Era, sometimes referred to as the "Age of Mammals," includes two periods, the Tertiary and Quaternary. Because the epochs of the Tertiary are characterized by very distinctive faunas, we will consider them separately. In the Paleocene Epoch (54–65 million years ago) a great flowering of mammals occurred, following the extinction of the dinosaurs. New Mexico Paleocene strata of the Nacimiento Formation are exposed as extensive badlands in the San Juan Basin, and contain some of the best and most abundant fossils of primitive mammals in the world. First discovered by David Baldwin in 1879, these fossils were initially studied by E. D. Cope in the 1880s, and the primitive nature of the mammals eventually led to the acceptance of the Paleocene as a new epoch of geologic time. Many additional collections of Paleocene vertebrates have been made by numerous institutions from the 1890s to the present, and specimens continue to erode out of the Nacimiento Formation. University of New Mexico and Bureau of Land Management paleontologists have collected hundreds of teeth and jaw fragments in the past few years, and many of these have added significantly to our knowledge of Paleocene life in general, and to our information on specific aspects of the skeletons of particular species. The international reference localities for early and middle Paleocene continental deposits are in the New Mexico part of the San Juan Basin, and the late Paleocene is represented by mammals from localities only a few miles north of the New Mexico–Colorado border.

The presence of well-preserved, abundant Paleocene fossils in rocks immediately above the fossiliferous Late Cretaceous formations makes the San Juan Basin sequence one of only a few in the world where the change from dinosaur-dominated to mammal-dominated terrestrial communities is excellently documented. The presence of coal in the Late Cretaceous Fruitland Formation, and plans to extract it by large-scale strip-mining operations, has led to concern by paleontologists about the effects mining and related activities will have on the Late Cretaceous–early Tertiary fossils of the region. The possible loss of scientific information about the life that existed during this time (new types of animals are continually being discovered) must be balanced against the advantages of developing New Mexico's coal resources for future energy needs. One hopes that procedures will be adopted that will ensure that this world-famous fossil field is not significantly damaged.

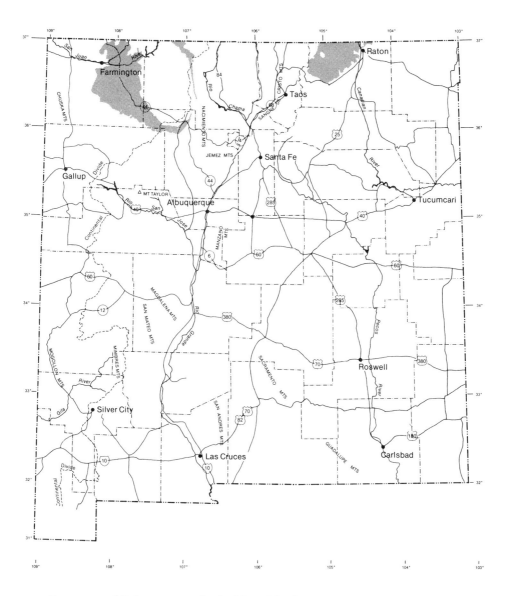

Outcrops of Paleocene rocks in New Mexico.

THE NEW MEXICO FOSSIL RECORD

Part of the jaw of *Allognathosuchus*, a crocodilian with blunt rounded teeth used to crush molluscs and/or turtles. Nacimiento Formation, San Juan Basin; about natural size.

Fossils present in the Nacimiento Formation include mammals, turtles, crocodilians, other reptiles, garfish, freshwater molluscs, and plants. The reptile fossils, primarily shell and bone fragments and teeth, are generally similar to those found in the underlying Late Cretaceous deposits. Several species of turtles are known, mainly from shell fragments, and four species of small- to medium-sized crocodilians have been described. Among the crocodilians are *Allognathosuchus*, which besides having normal conical teeth also possessed low, rounded teeth probably used for crushing molluscs or turtles. *Akanthosuchus*, just described in 1981, had bony scutes with sharp spikes and blades on its back, which would have given it a spiny appearance. In addition, a completely articulated hind limb of this crocodilian was discovered, after paleontologist Mike O'Neill, noting a single bone going back into a sandstone layer, chopped out a large block of the rock, took it back to the laboratory, and scraped away the matrix. In many cases, the single bone observed on the surface might have been all that was present, but the discovery of good fossils is often a matter of luck and educated guesses, and in this case the suspicions of the paleontologist that more might be present within the rock led to the rest of the foot coming to light, a unique and important find.

Though turtle and crocodilian fossils are common in the Nacimiento, other reptiles are rare. Small bones and teeth of lizards are occasionally found, and the distinctive spool-shaped vertebrae of champsosaurs, a group of aquatic reptiles resembling crocodilians but more closely related to lizards, have been reported. Garfish scales and bowfin vertebrae are relatively common fish remains. Several types of pelecypods and gastropods also lived in the freshwater environments; they are very similar to those found

PALEOCENE

An unusual articulated hind limb of another crocodilian, *Akanthosuchus*, from the Nacimiento Formation of the San Juan Basin. About natural size.

in the underlying Cretaceous beds. An unusual but indirect piece of evidence for the presence of ants is a petrified log containing galleries similar to those made by the modern carpenter ant, found near the Cuba-Bloomfield highway in the 1930s.

The relatively few plants that have been found in the Nacimiento Formation are mainly angiosperm leaves. They resemble the plants found in the Cretaceous, suggesting, at least in New Mexico, that the Cretaceous forest did not change greatly into the Paleocene. However, because the Cretaceous sea had retreated to the north far from New Mexico, the sediments of the Nacimiento Formation were deposited in continental basins—lowlands with numerous streams, rivers and perhaps lakes that supported subtropical forests somewhat less diverse than the luxuriant Cretaceous forests of this area.

Thousands of mammal fossils, mainly teeth and jaw fragments, but occasionally skulls and partial skeletons, have been collected

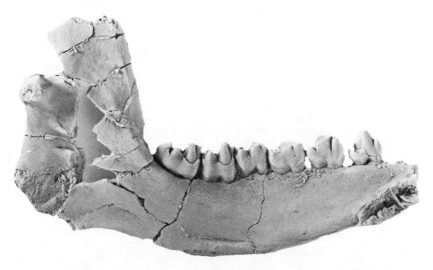

A small mammal jaw *(Gillisonchus)* from the Nacimiento Formation of northwestern New Mexico (original about 5 cm long). Most mammal remains are minute isolated teeth or small jaw fragments (the jaw illustrated was painstakingly pieced together from numerous fragments), but thousands of such fossils have been collected from the Nacimiento over the years. Because the mammal remains are so small, paleontologists collect by taking large samples of sediment and then screening it through sieves that concentrate the fossils.

from the Nacimiento Formation, and they tell us a great deal about the nature of the mammalian faunas living in Paleocene time. Paleocene mammals are primarily primitive variations of the generalized stock represented by the order Insectivora, which includes the modern shrews, moles, and hedgehogs. Most of the mammals belong to extinct orders, such as the creodonts and condylarths, from which more advanced mammalian orders later evolved. Creodonts, for example, were small carnivores with more primitive teeth and less brain capacity than the advanced carnivores that evolved from them and that today include cats, dogs, raccoons, weasels, and many other forms. Similarly, condylarths were primitive mammals whose generalized teeth show that many were becoming predominantly herbivorous. From condylarths several highly successful modern ungulate (hoofed mammal) orders arose, particularly the perissodactyls (horses, rhinoceroses) and

PALEOCENE

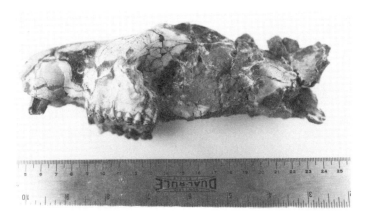

Occasionally, nearly complete skulls are found in the Nacimiento Formation. This specimen (of *Claenodon*, a relatively large condylarth) was found in 1977 near Lybrook, and is one of the best skulls of this genus known.

artiodactyls (pigs, sheep, cattle, deer, camels, etc.). Though ungulates are greatly different from carnivores today (there is no difficulty in distinguishing a cat from a camel!), their Paleocene creodont-condylarth ancestors were so similar in many respects that it is difficult to decide exactly which group some Paleocene fossils belong to.

A few very small, very primitive primate remains are also found in the New Mexico Paleocene; these are among the earliest representatives of the order that was eventually to include monkeys, apes, and man. These Paleocene primates, however, would have appeared more like modern tree shrews than any of the familiar advanced primates, and careful examination of their teeth is sometimes necessary to separate them from their insectivore forebears. In general, the Paleocene mammal fauna includes generalized primitive forms whose subsequent evolution through the Cenozoic led to the development of numerous advanced and highly dissimilar modern orders.

Primitive forms unrelated to any modern mammals were significant, too, particularly the multituberculates. This interesting and long-lived group (its range is from Jurassic to Oligocene; Cretaceous as well as Paleocene specimens have been recovered from the San Juan Basin) probably functioned as rodents do today; in fact, their extinction is thought to be due to the rise of rodents in the Eocene. The teeth of multituberculates are very distinctive, consisting of long rodentlike incisors, semicircular serrated premolars, and molars that have two or three long parallel rows of numerous cusps—totally unlike the teeth of any other mammal that has ever lived.

The Paleocene mammal fossils of the San Juan Basin indicate a rapid radiation and diversification immediately following the extinction of the dinosaurs. Though several dozen different species (some of them unique to New Mexico) lived in the Basin, they were all relatively small, from the size of modern shrews to that of a large dog. Not enough time had elapsed in the Paleocene to allow really large mammals to evolve from their small Cretaceous ancestors, but evolutionary trends in some groups were already pointing in that direction.

In northeastern New Mexico, around Raton and north into the Trinidad area of Colorado, Paleocene exposures have yielded impressive plant fossils. These plants are highly diverse (nearly 150 species have been reported), very abundant, and well preserved in some beds, and have been extensively carbonized into the coal deposits that are mined in this area today. In general, the plants are similar to those found across the state in the San Juan Basin (but more diverse and better preserved), and they indicate the presence of large forests of deciduous trees, with smaller amounts of conifers and ferns. The environment of at least part of the Raton area during the Paleocene was probably marshy or swampy, with waterlilies, ferns, and other plants living in or near shallow ponds. Around the marshes grew trees with leaves that resemble those of modern palms, figs, cinnamons, breadfruits, magnolias, sweetgums, and sycamores. Several types of climbing vines adorned these dense forests. On higher and drier ground, trees that may be related to modern oaks, walnuts, beeches, chestnuts, lindens, and laurels grew. The climate indicated by these diverse floras is warm-temperate to humid-subtropical, perhaps similar to the climate of modern-day Georgia or South Carolina. Unlike the Nacimiento Paleocene deposits, however, fossils of animals are

Restoration of *Ptilodus,* a member of the multituberculates, from the Paleocene Nacimiento Formation. Animal was about 30 cm long. *(After Kurtèn)*

virtually nonexistent in the Raton beds. Where were all the animals? Were they originally present in the forests and not preserved as fossils for some reason? Or was there something about the forest environment of northeastern New Mexico that made it unsuitable for many animals to live in? Paleontologists have no good answer yet.

Eocene

The Eocene Epoch (37–54 million years ago) is represented in New Mexico by a large area of red, maroon, and brown shales and sandstones in the center of the San Juan Basin (San Jose Formation), and by restricted mostly sandstone outcrops between Albuquerque and Sante Fe (Galisteo Formation), in the Datil area and southeast of Socorro (Baca Formation), and in the Rincon Mountains in southern New Mexico. By far the most abundant and diverse Eocene fossils are in the early Eocene San Jose Formation, but some late Eocene fossils have been found in the Galisteo and Baca formations.

The discovery and description of Eocene vertebrates in the Almagre and Blanco Arroyo systems west of Gallina by E. D. Cope in 1874 was one of this great vertebrate paleontologist's most significant contributions. The discovery was almost not made. Cope was a member of the Wheeler Geological Survey, whose leader had no plans for fossil-collecting in the area. Cope, unable to resist the lure of the badlands, set off on his own to explore them, later rejoining the expedition in Tierra Amarilla. In his own words:

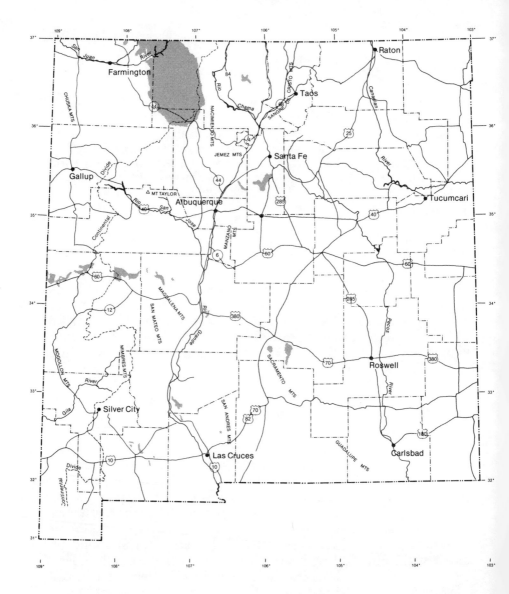

Outcrops of Eocene rocks in New Mexico.

EOCENE

As soon as we picketed the horses, we began to find fossil bone! The first thing was a turtle and then *Bathmodon* [now *Coryphodon*] teeth! and then everything else rare and strange till by sundown I had 20 species of Vertebrates! all of the lowest Eocene . . . The most important find in geology I ever made . . .

Cope's enthusiasm echoes through the years and conveys the delight every paleontologist experiences when he has found a richly fossiliferous deposit filled with unusual or previously unknown kinds of fossils. Subsequent collecting from the Eocene in the San Juan Basin has increased the number of different species known from the San Jose Formation, and it stands as one of the best deposits of early Eocene fossils in the country.

As in the Cretaceous and Paleocene strata discussed previously, crocodilian, turtle, and garfish remains are common in the San Jose. Snakes and lizards, as usual, are very rare. Plants and freshwater molluscs are uncommon, but in 1980 a good bed of well-preserved angiosperm leaves was discovered that will add considerably to our knowledge of Eocene paleobotany in New Mexico. Brightly colored petrified logs and stumps are present in some sandstone beds of the San Jose. The most important fossils are mammals, and they indicate that several notable advances in mammalian evolution had occurred by early Eocene time.

The majority of fossils are of the archaic mammal groups that had dominated the Paleocene—insectivores, the carnivorous creodonts, and the herbivorous/omnivorous condylarths. Members of two other orders, the tillodonts and taeniodonts, are also present, but these groups became extinct at the end of the epoch. Some of these primitive mammals reached much larger sizes than their Paleocene ancestors. One of the most common San Jose fossils is *Coryphodon*, an archaic ungulate with a massive body and the proportions of a small hippopotamus. *Coryphodon* is thought to have been an aquatic herbivore, and the discovery in the 1940s of a concentration of *Coryphodon* skeletons in north-central New Mexico suggests that a group of them perished as a pond dried up. This unusual assemblage of these giants of the early Eocene has been studied by Spencer Lucas, this country's expert on New Mexico Eocene fossils.

In addition to the extinct mammalian groups mentioned above, the New Mexico early Eocene also includes some of the earliest representatives of advanced orders that are still with us today. A fairly common type is *Hyracotherium* (once called *Eohippus*), an animal about the size of a small dog, which was the first horse.

Fragment of a turtle shell. This is an Eocene fossil, but similar fragments are common in Late Cretaceous and Paleocene rocks in the San Juan Basin. About one-half natural size.

Hyracotherium lived in forests and ate leaves; it had four toes on the front feet and three on the hind ones. From this humble beginning horses increased in size, reduced the number of toes ultimately to one on each foot, and developed large flat teeth with complicated ridges as they adopted a grazing rather than a browsing existence. The progressive changes shown by horses from the Eocene to the present are abundantly documented in the fossil record, and provide an excellent example of evolution within a single group over an extended period of time.

Also found in the early Eocene of the San Juan Basin are fossils of small true carnivores (miacids) that would later begin to diversify into the various modern carnivore groups (dogs, cats, weasels, and so forth). Early rodents, small squirrellike gnawing mammals, are also present in the San Jose, and likewise represent a group that would become increasingly successful as time went on. Rare

Coryphodon jaw, from the early Eocene San Jose Formation north of Cuba. About one-half natural size.

specimens of even-toed ungulates (artiodactyls) and primates—mostly teeth—have been recovered from several localities. Other important mammal orders that begin in the Eocene (but that have not been found in New Mexico) are such diverse but familiar forms as whales and elephants. We can consider the Eocene of New Mexico as still dominated by archaic mammal groups, but with a conspicuous component of primitive representatives of many advanced, modern orders as well.

We tend to forget, because bird fossils are so much rarer than mammal fossils, that the early Tertiary was a time when birds were becoming successful, and perhaps were competing with mammals for dominance of terrestrial environments. Only one type of bird is known from the New Mexico Eocene, but it certainly seems capable of having wreaked havoc among the local mammals. This is *Diatryma*, the "terror crane," a huge flightless

Top: *Coryphodon*, an archaic ungulate about the size of a cow or small hippopotamus. *(After Kurten)* Bottom: Two Eocene condylarths, *Phenacodus* (about two meters long), and the small *Meniscotherium*, both from the San Jose Formation of New Mexico. These two genera illustrate the diversity within this primitive but important group of early Tertiary mammals. *(After Kurten)*

Jaw fragment of *Hyracotherium*, the first horse. About natural size.

Restoration of *Hyracotherium*, about the size of a small dog. *(From a painting by Z. Burian)*

bird more than two meters tall with a large head and thick powerful beak. *Diatryma* and related forms perhaps represented an early attempt by birds to fill the large-carnivore niches left open after the extinction of the dinosaurs. These large birds declined rapidly, however, as more advanced and intelligent mammalian carnivores developed.

Though most Eocene fossils in New Mexico have come from the San Jose Formation (early Eocene), exposures of the Galisteo Formation between Santa Fe and Albuquerque have yielded some very interesting late Eocene mammals. A thin bonebed exposed in an arroyo in this area is composed largely of bones of titanotheres, a group of extinct, horned, odd-toed ungulates related to horses that reached the size of small elephants. Some of the teeth found in the Galisteo are seven cm long and have the raised W-shaped ridge patterns characteristic of titanotheres. Also in the

The large flightless bird, *Diatryma*, more than two meters high. *(From a painting by Z. Burian)*

late Eocene part of the Galisteo are the fragmentary remains of primitive rhinoceroses, artiodactyls, carnivores, and rodents, but these are very rare. Petrified wood is abundant, however; apparently pines, oaks, beeches, and poplars grew in profusion in this area. Petrified logs over a meter in diameter and 50 meters long were once common, but these fossils have become severely depleted in the past few decades because of souvenir hunters. A note of historical interest: this petrified wood is the first type of fossil ever reported from New Mexico. Josiah Gregg, a trader on the Santa Fe Trail, noted the brightly colored wood in the 1830s, and mentioned it in his diary, published in 1844.

Oligocene

Few valid occurrences of Oligocene (25–37 million years ago) fossils have been reported from New Mexico. A plant locality in the San Mateo Mountains contains mostly deciduous tree leaves and some pines, suggestive of a temperate climate. Near Hillsboro and Hermosa numerous Oligocene fossils of an extinct member of the bristlecone pine group have been found in lake beds associated with a volcanic sequence. Some vertebrate fossils from sandstone exposures (Baca Formation) in west-central New Mexico, particularly in the Datil area, have been interpreted as Oligocene rather than late Eocene, but the exact age has not been settled to everyone's satisfaction. These fossils include titanotheres similar to those in the Galisteo Formation, as well as artiodactyls and creodont carnivores, but the fossils are generally very fragmentary and sparse.

Bottom of the skull of *Teleodus*, a large brontothere from the Galisteo Formation near San Ysidro. The teeth show the W-shaped ridges characteristic of brontotheres. About one-fifth natural size. Specimen in American Museum of Natural History.

Miocene-Pliocene

In 1874, about three weeks before he discovered early Eocene fossils in the Gallina area, Cope traversed the badlands around San Ildefonso and Pojoaque Pueblos north of Santa Fe. He collected numerous fragmentary fossils of geologically young vertebrates and soon described several new species, but never returned to follow up his discovery. For fifty years after Cope's brief explorations, these badlands lay exposed to the erosive forces of wind and water, revealing multitudes of fossil bones and skeletons, a scientific bonanza to whatever paleontologist would take a second look at this barren landscape.

Finally, in 1924, Childs Frick, an independently wealthy paleontologist working in association with the American Museum of Natural History in New York, sponsored an exploratory expe-

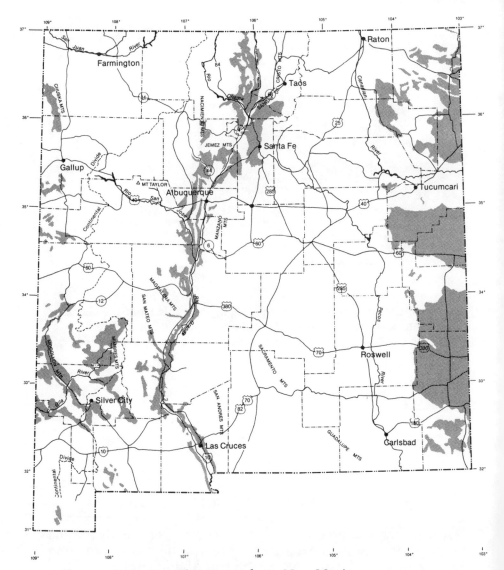

Outcrops of Miocene-Pliocene rocks in New Mexico.

dition to the Santa Fe badlands. Within a week the field party had found a complete skeleton of the "dog-bear" *Hemicyon,* and the remains of other vertebrates proved to be so varied and abundant that Frick began a collecting program that ran almost continually during summers for the next forty years, ending in the 1960s. The quantity of vertebrate fossils removed during this time was staggering—thousands of individual bones and about 200 more-or-less-complete skeletons, representing well over 100 different kinds of animals. This concentrated collecting effort, unique in the history of New Mexico paleontology, established the Santa Fe Group as having one of the most abundant, diverse, and continuously sampled late Tertiary faunas in the world. Study of the fossils showed that they represent faunas of middle Miocene to early Pliocene age (about 5 to 18 million years old). Because the Santa Fe faunas extend through portions of the Miocene and Pliocene Epochs, we will consider the two epochs together.

Santa Fe Group fossils are almost entirely mammals. They include a mixture of essentially modern forms, primitive representatives of some modern groups, and unfamiliar mammals that are now extinct. A few reptiles, plants, and birds are also present. Camels are the most abundant mammals; there were at least twelve species, ranging from forms the size of gazelles to some that were close in size and proportions to giraffes. Primitive deer, deerlike animals, and antelopes were also abundant. A wide variety of horn types characterize these mammals, and pieces of horns are among the most frequently found of their fossils. The horses of the Santa Fe Group were more similar to modern horses than to their small Eocene ancestors, but were somewhat smaller than modern horses and possessed three toes on each foot. Oreodonts were a very successful group of grazing animals, perhaps best visualized as having a combination of pig and small camel features. Although oreodonts were once the most abundant mammals in North America, roaming in vast herds over the Midwest in the Oligocene and Miocene, by the time Santa Fe Group deposition began they had passed the peak of their success, and they are but a small part of the vertebrate fauna in New Mexico. Some of the last surviving members of this extinct group are found in the Pliocene part of the Sante Fe.

The largest of the Santa Fe mammals were the gomphotheres, extinct predecessors of modern elephants that possessed tusks in the lower as well as upper jaw. Several types of rhinoceros also are present in the Sante Fe, including *Teleoceras,* a presumably amphibious genus with short legs and a stout body reminiscent

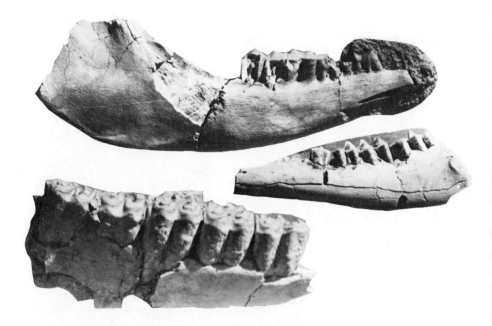

Top: Partial jaw of a rhinoceros from the Santa Fe Group near Española. About one-half natural size. Center: Partial jaw of a merycodont (deerlike animal) from the Santa Fe Group. About natural size. Bottom: Lower jaw fragment of the Miocene horse *Merychippus* from the Santa Fe Group, near Española. Five complexly ridged grinding teeth are visible. Photo is about natural size; horses had evolved considerably by Miocene time from the small creature of the Eocene.

of the modern hippopotamus. Other herbivorous mammals present in the Santa Fe Group are rabbits, beavers and other rodents, peccaries, ground sloths, shrews, moles, and hedgehogs.

Carnivorous mammals are not as abundant as the large herbivores discussed above, but are highly diverse. Some forms were similar to modern dogs, foxes, cats, raccoons, badgers, skunks, and weasels, but groups now extinct were present as well. Most Santa Fe cats, for example, possessed enlarged canine teeth that would qualify them as "sabre-toothed," and there was a large variety of bulky doglike carnivores, some of which were probably close to the ancestry of bears.

Other vertebrates are much less common than mammals in the Santa Fe Group. One genus of vulturelike bird is present, and several turtles and lizards have been reported. The nearly complete shell of a large land tortoise, measuring almost a meter in

Top: Well-preserved skull of a doglike carnivore from the Santa Fe Group north of Santa Fe. Fossils like this attest to the fact that even after 40 years of intensive collecting, excellent remains continue to erode out of Santa Fe Group rocks. About natural size. Bottom: Land tortoise shell from Santa Fe Group near Española. Original is about one meter long.

length, was found in the 1930s in Santa Fe badlands around the Santa Cruz–Nambe area. This specimen, which has part of its shell crushed (perhaps the result of being stepped on by a gomphothere), is on exhibit in the University of New Mexico Geology Museum. An unusual occurrence of fossil amphibians is marked by the fragmentary remains of a frog. Aside from rather small amounts of unidentifiable petrified wood, the only good plant fossil is a palm frond.

The wonderful fossil record in the Santa Fe Group allows us to visualize the environment these organisms lived in with great accuracy. Much of the Santa Fe area consisted of savannahlike plains in areas of low relief, with more luxurious vegetation along streams, and forests in higher elevations. Herds of horses, camels, "antelope," rhinoceroses, and oreodonts would have been visible on the plains along with cats, dogs, and tortoises, while local forested areas would have been the habitat of gomphotheres, some "deer," and some rhinos. Along the streams, beavers, some rodents, shrews, and smaller carnivores like skunks, raccoons, and weasels would have lived. The climate was warm and probably humid, with mild, possibly frost-free winters. As might be expected from the intensive collecting effort by the American Museum, the Santa Fe Group is not as rich in exposed fossils as it once was, but pieces of bone and occasionally teeth and jaws may still be found. Some of the more spectacular Santa Fe fossils are on exhibit at the American Museum in New York City.

Although typical Santa Fe deposits are restricted to the Santa Fe–Española–Abiquiu area, sediments of similar lithology but usually younger ages are found all along the Rio Grande and in many places in southwestern New Mexico. These deposits are sparsely fossiliferous, but occasional mammals are found in them, though these are almost always small skeletal fragments. In addition, opalized wood in many colors, some of it of gem quality, has been found in many areas along the Rio Grande, including the north end of the Fra Cristobal Range south of San Marcial, and an area between Los Lunas and Socorro.

All along the High Plains region of eastern New Mexico, and into the Texas panhandle, sediments of Pliocene to possibly Pleistocene age (Ogallala Formation) are exposed. Numerous Pliocene mammals have been found in Texas, but so far very few vertebrate remains have been reported from the New Mexico Ogallala, though some molluscs and plant fossils have been described. These are very much like modern types, and a fuller discussion of them will be found in the section on the Quaternary of New Mexico.

Quaternary

The Quaternary Period (about 2 million years ago to the present) includes the Pleistocene and Recent Epochs, the latter beginning only about 10,000 years ago. During the Pleistocene at least four waves of glaciers advanced down from the north, covering most of Europe and much of North America with thick ice sheets. The glaciers did not reach as far south as New Mexico, but small mountain glaciers formed in some of the state's mountain ranges. North American Pleistocene faunas were characterized by numerous large mammals, many of which became extinct at the end of the epoch as the last glaciers retreated, leaving a modern

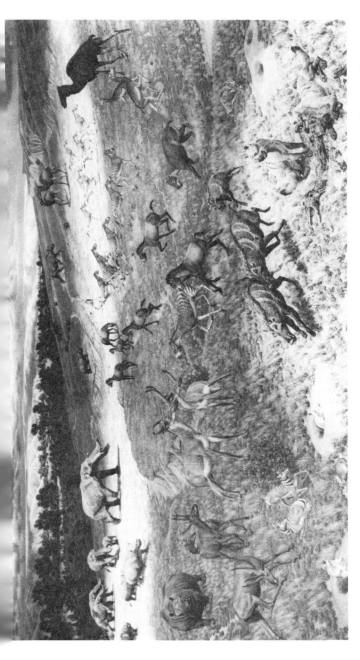

An early Pliocene landscape and mammals. Though this scene reconstructs conditions in the area from Texas to Nebraska, the vegetation and many of the mammals shown are also representative of the early Pliocene in north-central New Mexico during the time of Santa Fe Group deposition. Rhinoceroses such as *Teleoceras* (far left) and *Aphelops* (center background), camels (upper right), three-toed horses (center), deerlike ungulates with two horns above the eyes and a third projecting from the back of the head (*Cranioceras*, far left foreground), the "dog-bear" *Hemicyon* (center right, chasing horses), and the short-faced dog *Osteoborus* (lower right) all lived in New Mexico during Pliocene time. *(From a mural by Jay Matternes; reproduced with permission of the artist and the Smithsonian Institution)*

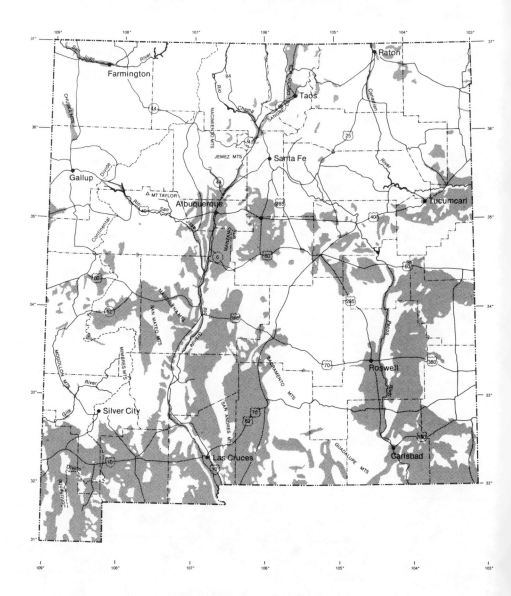

Outcrops of Quaternary age in New Mexico.

QUATERNARY

fauna that is severely depleted of many types of mammals that were abundant and successful during the Pleistocene. Also during the Pleistocene *Homo sapiens,* or modern man, evolved, and his effects on modern faunas and floras have been the dominant factor in the progression of life from about 10,000 years ago to the present.

Large areas of poorly consolidated sediments and alluvium of Quaternary age are exposed in most parts of New Mexico, often as gravelly river channel and bank deposits or fine-grained lake bed sediments. Much volcanic activity occurred in the state during the Quaternary, leaving large areas of lava and ash associated with volcanic cones, and in the Jemez area, with the Valles Caldera, an enormous collapsed volcano.

Pleistocene vertebrates are widespread in New Mexico sediments; about a hundred localities have been reported. At most sites, however, the vertebrate fossils are scattered, fragmentary, and rarely abundant. By far the largest number of Pleistocene vertebrates have come from sites of late Pleistocene age, 10,000 to 20,000 years old, and in some of these (Folsom, Blackwater Draw) the remains are mixed with human artifacts, indicating that prehistoric Paleo-Indians killed the animals and used them for food. In addition to their occurrence in river gravels (such as around Albuquerque and near Las Cruces), lake beds (San Augustin Plains), and other sediments, several caves have yielded good vertebrate remains. Isleta Cave (south of Albuquerque), Sandia Cave (Sandia Mountains), and several caves in the Guadalupe, Little Hatchet, and Organ mountains of southern New Mexico have all produced Pleistocene vertebrate fossils. From these occurrences we have a fairly good idea of the nature of Pleistocene faunas in New Mexico.

New Mexico Pleistocene vertebrates include many mammals that are still living in the state, plus a significant component of large mammals that became extinct within the last 20,000 years. Among the most abundant of the extinct forms are mammoths and mastodons, large relatives of the modern elephant. Although their skeletons and appearance were similar, mammoths and mastodons are easily distinguished by their teeth, which are among the most commonly found of their fossil remains. Mammoth teeth are up to 30 or 40 cm long and have numerous distinctive ridges aligned perpendicular to the length of the tooth on the grinding surface, that allowed them efficiently to grind up the grasses that formed the main part of their food. Mastodon teeth are about the same size as mammoth teeth, but have sev-

THE NEW MEXICO FOSSIL RECORD

Part of a mammoth jaw, showing the grinding surface of a single large molar tooth. The tooth is about 25 cm long. From east of Capitan Mountain, New Mexico. Mammoth teeth and tusk fragments have been found occasionally in many parts of the state.

eral rounded cusps, indicating that they browsed on soft leaves instead of grass, and therefore lived in forests. Large curved tusks, or pieces of them, along with parts of the postcranial skeleton of these large beasts have also been found.

Several species of extinct horses also lived in New Mexico during the Pleistocene; these were essentially similar to modern wild horses and the African zebra. Horses became extinct in North America about 10,000 or 15,000 years ago, but were reintroduced by the Spaniards during the 1500s. Camels and llamas were conspicuous Pleistocene inhabitants of the state, too. A beautifully preserved, virtually complete skeleton of a young camel was recently excavated from a gravel pit north of Albuquerque, where it had been exposed by a bulldozer. It is one of the best Pleistocene vertebrate fossils that has been recovered in New Mexico.

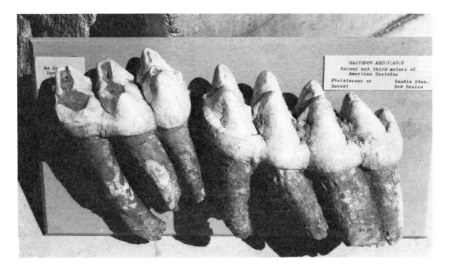

Mastodon molars from the Sandia Mountains. Compare with the mammoth molar; though these two groups of elephantlike mammals resembled each other closely in their external appearance, their teeth are quite different.

Other large herbivorous mammals include extinct muskoxen, caribou, and several types of pronghorn "antelope." Herds of bison roamed New Mexico during Pleistocene time, and remains of these animals, some of which were much larger than modern buffalos, are among the most common of the state's extinct Pleistocene mammals. Ground sloths, large relatives of the modern tree-sloths of Central and South America, also lived in New Mexico, but their fossils are relatively uncommon. One specimen, however, of a medium-sized (two-meter-long) sloth found in the 1920s, is remarkably well preserved. Apparently it had fallen into a volcanic fissure near Aden Crater, southwest of Las Cruces, about 11,000 years ago and had become mummified in the dry desert air. When found by some youths, not only the skeleton was intact but also its fur and some elements of its soft anatomy

Restoration of the "Columbian mammoth," and *Smilodon*, the sabre-toothed cat, both known from remains at Blackwater Draw. *(From a painting by Z. Burian)*

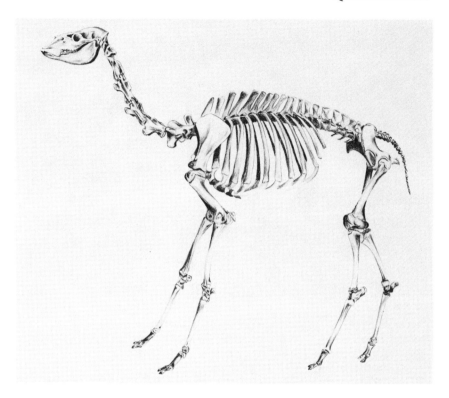

Skeleton of *Camelops*, a Pleistocene camel excavated in 1979 from a gravel pit in Albuquerque (see frontispiece).

were preserved as well. Its coprolites (fossil fecal material) were also studied and the partially digested plants allowed a determination of what the sloth had eaten for its last meal.

Carnivorous mammals, as usual, are less common than herbivores, but cave bears, dire wolves, large lions, and sabre-toothed cats are among the extinct carnivores that preyed upon the large herbivorous mammals. Dire wolves were considerably larger than modern wolves; similarly, the lions were bulkier than their surviving African relatives. The sabre-toothed cat, so called because its two upper canine teeth were elongated into 15-cm-long curved daggers, was a formidable predator that probably fed mainly on mammoths and other larger herbivores, using its "sabres" to stab the animal quickly to death. The depletion of large mammals at the end of the Pleistocene caused the extinction of sabre-toothed cats, whereas other cats, with shorter canines, but more varied food preferences, survived to the present.

THE NEW MEXICO FOSSIL RECORD

Restoration of the ground sloth *Nothrotheriops* (formerly *Nothrotherium*) and *Smilodon*, immediately before the sloth stumbled into the volcanic fissure near Aden Crater, eventually to become mummified. *(From a painting by Z. Burian)*

In several New Mexico Pleistocene fossil localities the remains of extinct mammals are found in association with the artifacts of Paleo-Indian hunters. The Folsom site in northeastern New Mexico is one of the best-known occurrences; here, spear points were found embedded in the skeletons of giant bison, and it is indisputable that these animals were being killed and used by the Paleo-Indians for food. The discovery of the Folsom site in the 1920s necessitated a full-scale reevaluation of the antiquity of man in North America, pushing the date of man's entrance into North America back much farther than previously had been believed possible. Soon afterward, in the 1930s, study of the Blackwater Draw localities between Clovis and Portales provided additional evidence of the coexistence of Paleo-Indians with a wide variety of extinct Pleistocene mammals. Here, mammoths were apparently the main food item; several of them show signs of

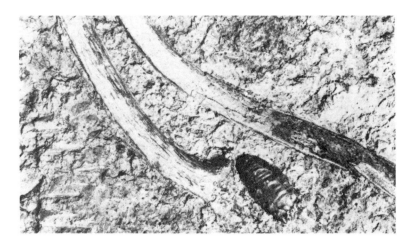

Paleo-Indian spear point found embedded between the ribs of an extinct bison at the Folsom site in northeastern New Mexico.

having been butchered on the spot at one locality. Study of the Blackwater Draw faunas has continued to the present, and its standing as one of the most important Pleistocene mammal/human sites in the United States is assured.

Radiocarbon dating indicates the Folsom, Blackwater Draw, and similar sites to be between 10,000 and 12,000 years old. It was during or shortly after this time that camels, horses, mammoths, ground sloths, and other large Pleistocene mammals—groups that were common and successful inhabitants of North America for a long time previously—completely vanished from this continent, though modern representatives still live elsewhere. Some paleontologists argue that man was primarily responsible for these extinctions; that development of an efficient hunting technology, man's fast migration through North America, and the large mammals' lack of prior experience with such a potent predator all combined to assure their termination in a surprisingly short time. Others believe that significant climatic changes accompanying the last retreat of the glaciers were primarily responsible for the extinctions. The precise causes are currently the subject of a vigorous debate; perhaps human and climatic effects acting together is the best answer. Whatever the explanation, since about 8,000 years ago, only yesterday in terms of geologic time, no trumpeting of mammoths, rhythm of wild horse's hooves, or braying of native camels has penetrated the clear New Mexico air, and the mammals around us today are but a smaller shadow of the faunas that once lived here.

In addition to mammals, some of the caves in southern New Mexico have yielded fossils of dozens of different birds. Most of these birds are species that are still living but several extinct species of birds are known as well. These include an extinct roadrunner, turkey, eagle, and vulture. In addition, birds very similar to the modern giant condor once soared through the skies over the state.

Shells of Pleistocene pelecypods and gastropods have been found in more than 100 localities, primarily in the eastern and southern parts of New Mexico. Nearly all of the dozens of reported species are small forms that may still be found living in freshwater or terrestrial environments around the Southwest. Modern populations of many of these species, however, now live in different areas than their Pleistocene ancestors, indicating that significant changes in New Mexico's climate and environments have occurred in the past few thousand years. Some species that lived in New Mexico during the Pleistocene can no longer be found within the state. Detailed study of Pleistocene molluscs and mammals that have survived to the present, and a knowledge of the ecological preferences of modern representatives of various Pleistocene species have allowed paleontologists and biologists to understand better the climatic changes that have affected New Mexico in the recent past.

The Present

Description of the modern fauna and flora of New Mexico falls within the realm of biology; the work of the paleontologist merges with that of the biologist as modern organisms prevail and the present is closely approached. The panorama of long-vanished life forms preserved as fossils in the state's rocks, however, brings

THE PRESENT

home the realization that the animals and plants we see around us today are only transitory parts of the great changing continuum of life. The tremendous variety of organisms that have lived in New Mexico during the past 500 million years informs us, too, of the myriad changes that have occurred in the climate, landscapes, and environments that have preceded and yet contributed to the New Mexico we know today. The uniformitarian dictum that "the present is the key to the past" can easily be turned around: knowledge of the past is essential to an understanding of the present. A species of animal, or a mountain range, or a human city are all products of past events, and can only truly be comprehended if the processes, forces, and events that influenced their development are known. And only with a knowledge of the past can we extrapolate the present into the future.

It has been the fate of the great majority of organisms to become extinct. Should we worry, then, that mankind has directly caused the extinction of hundreds of species in its short time on earth? All the organisms we see in the fossil record attained success only upon becoming well adapted to their environments. Environmental changes generally cause extinction, or in a minority of cases, the evolution of one species into another better adapted to the new environment. The environment in which we humans live has undergone enormous changes in the past few thousand years; many of these changes are of our own making and some are intensifying yearly. What are the implications of these changes for the future of our own species? Sixty-five million years ago an asteroid impact may have destroyed one-half or more of the different kinds of organisms then living. What would be the effects on present life, including our own species, if a similar impact occurred tomorrow? An understanding of the history of life, besides being fascinating in itself, can lead the contemplative individual into a consideration of our future, and the future of the other life forms with which we share this planet.

The Late Cretaceous ammonoid *Coilopoceras*, exposed in a concretion near the Rio Puerco, west of Albuquerque.

The Ammonite

I am the ammonite.
I grew my own spiralled shell and swam in Mesozoic seas.

It's all gone now.
I sit in my glass case and watch the shadows peer,
Look at their watches, and disappear.

We had it good in those old days,
Swimming and eating our way through a hundred and sixty
   million years
Before things started to get tough.
The big-footed eaters died out.
We all died out.

Buried in stone, I waited.
Seventy million years passed like a dream.
Then the stone parted, washed away to the growing light,
And I looked out on a dry world.

A shadow came and took me away.
He drilled a hole through my shell and wore me around his neck.
I witnessed strange rituals;
The power of my years was called upon to heal and protect.

He died and I was buried again.
Two thousand years passed like a flicker.
Then another shadow dug me up and put me in this case
To watch the shadows come and go.

It's been a quick thirty years;
Hardly enough time to catch my breath and wonder what
   to expect
In the next million years.

                                                         Ted Greer

WHERE TO GO FOR FURTHER INFORMATION

In General

Though this book provides an introduction to New Mexico's fossils, some readers may encounter fossils that are not discussed or illustrated in it, or may desire to extend their knowledge beyond the scope of the book's coverage. The references cited below provide more detailed, but in most cases still fairly general, information on various aspects of New Mexico's fossil record, and the interested reader is urged to peruse them. An excellent source of information about the geology and paleontology of specific parts of the state are the Field Conference Guidebooks published annually by the New Mexico Geological Society. Each guidebook covers a restricted part of the state, includes roadlogs that summarize the geology and other natural and historical features the traveler encounters on the roads of the area, and contains technical and general articles on various aspects of the area's geology, including paleontology. The New Mexico Bureau of Mines and Mineral Resources series "Scenic Trips To the Geologic Past" also provides general geological discussion, including brief mention of fossils, of selected parts of New Mexico.

Another source of information about New Mexico fossils are the paleontologists in the state; they are always happy to identify fossils that are brought to them for inspection and to talk with people interested in paleontology. Paleontologists are currently associated with the following New Mexico institutions: University of New Mexico Department of Geology (Albuquerque); New Mexico State University Department of Geology (Las Cruces); New Mexico Bureau of Mines and Mineral Resources (Socorro);

WHERE TO GO FOR FURTHER INFORMATION

Albuquerque and Farmington branches of the United States Bureau of Land Management; and the New Mexico Museum of Natural History (Albuquerque).

References

Ash, S. R. and Tidwell, W. D., 1976, "Upper Cretaceous and Paleocene Floras of the Raton Basin, Colorado and New Mexico." *New Mexico Geological Society Guidebook*, 27th Field Conference, pp. 197–203.
A summary, with some illustrations, of the plant fossils of the Raton area.

Cobban, W. A., 1977, "Characteristic Marine Molluscan Fossils from the Dakota Sandstone and Intertongued Mancos Shale, West-central New Mexico." U.S. Geological Survey Professional Paper 1009, 30 pp.
An illustrated guide to many of the characteristic mollusc fossils found in the widely exposed marine deposits of northwestern New Mexico.

Colbert, E. H., 1974, "The Triassic Paleontology of Ghost Ranch." *New Mexico Geological Society Guidebook*, 25th Field Conference, pp. 175–178.
A good summary of the geology, fossils, and environment during Triassic time in north-central New Mexico, around the site of one of New Mexico's most important fossil localities.

Colbert, E. H. and Northrop, S. A., eds, 1950, *Guidebook for the Fourth Field Conference of the Society of Vertebrate Paleontology in Northwestern New Mexico*. American Museum of Natural History and University of New Mexico, 91 pp.
Includes a roadlog and four articles that essentially summarize northwestern New Mexico vertebrate paleontology in the Paleozoic, Mesozoic, and Cenozoic Eras. Although a little outdated, this is an excellent introduction to the state's major deposits of fossil vertebrates.

Findley, J. S., Harris, A. H., Wilson, D. E., and Jones, C., 1975, *Mammals of New Mexico*. University of New Mexico Press, 360 pp.
Primarily concerned with modern mammals around the state, but an appendix comprehensively summarizes New Mexico's late Pleistocene mammals.

Flower, R. H., 1970, "Early Paleozoic of New Mexico and the El Paso Region." New Mexico Bureau of Mines & Mineral Resources, 44 pp.
The best available summary of the stratigraphy and paleontology of Cambrian through Devonian rocks in New Mexico.

REFERENCES

Kelley, V. C. and Northrop, S. A., 1975, *Geology of Sandia Mountains and Vicinity, New Mexico.* New Mexico Bureau of Mines & Mineral Resources Memoir 29, 135 pp.
Comprehensive treatment of the geology of the Sandia Mountains area, with tables and discussion of all of the types of fossils reported from this area.

Kues, B. S., 1979, "The Treasures of Paleontology." *El Palacio,* v. 85, no. 4, pp. 7–15, 30–33.
A nontechnical summary, with illustrations, of some of New Mexico's most important fossil deposits.

Kues, B. S. and Lucas, S. G., 1979, "Summary of the Paleontology of the Sante Fe Group (Mio-Pliocene), North-central New Mexico." *New Mexico Geological Society Guidebook,* 30th Field Conference, pp. 237–41.
General summary of the types of fossils found in the 5-to-18-million-year-old badlands between Santa Fe and Española, one of New Mexico's most prolific areas for mammal fossils.

Kues, B. S. and Northrop, S. A., 1981, *Bibliography of New Mexico Paleontology.* University of New Mexico Press, 150 pp.
Cites all technical and nontechnical publications dealing with New Mexico fossils.

Kurten, B., 1971, *The Age of Mammals.* Columbia University Press, 250 pp.
An excellent summary of mammal evolution through the Tertiary around the world. Includes discussion and a few illustrations of important constituents of New Mexico's major mammalian fossil deposits.

Laudon, L. R. and Bowsher, A. L., 1949, "Mississippian Formations of Southwestern New Mexico." *Geological Society of America Bulletin,* v. 60, pp. 1–88.
Detailed summary of the stratigraphy and paleontology of the Mississippian of southern New Mexico.

Lee, W. T. and Knowlton, F. H., 1917, *Geology and Paleontology of the Raton Mesa and Other Regions in Colorado and New Mexico.* U.S. Geological Survey Professional Paper 101, 450 pp.
This large monograph, though out of date in some places, provides information and fossil lists for more than 100 localities in northern New Mexico (it is not limited to the Raton area).

Lucas, S. G., 1977, "Vertebrate Paleontology of the San Jose Formation, East-central San Juan Basin, New Mexico." *New Mexico Geological Society Guidebook,* 28th Field Conference, pp. 221–25.
A summary of the vertebrate fossils of the Eocene San Jose Formation, one of the best Eocene fossil deposits in the United States.

Lucas, S. G., Rigby, J. K., Jr., and Kues, B. S., eds., 1981, *Advances in San Juan Basin Paleontology.* University of New Mexico Press, 393 pp.
Contains fifteen papers describing and summarizing the latest pale-

WHERE TO GO FOR FURTHER INFORMATION

ontological discoveries in the Cretaceous and early Tertiary of northwestern New Mexico.

Matthew, W. D., 1937, *Paleocene Faunas of the San Juan Basin, New Mexico.* Transactions of the American Philosophical Society, v. 30, 510 pp.

This magnificent monograph describes in detail the world-famous Paleocene vertebrate faunas of northwestern New Mexico.

Moore, R. C., ed., various dates, *Treatise on Invertebrate Paleontology.* Geological Society of America and the University of Kansas Press.

Issued over the years in about thirty volumes, the "Treatise" provides descriptions and illustrations of every genus of invertebrate fossil known. Most volumes also have much information on the evolution, distribution, and paleoecology of the specific group under consideration.

Moore, R. C., Lalicker, C. G., and Fischer, A. G., 1952, *Invertebrate Fossils.* McGraw-Hill Book Co., 766 pp.

A little outdated, but still the best introductory invertebrate paleontology textbook available. Covers all groups of invertebrates and is profusely illustrated.

Newell, N. D., et al., 1953, *The Permian Reef Complex of the Guadalupe Mountains Region, Texas and New Mexico.* W. H. Freeman & Co., 236 pp.

A comprehensive summary of the geology, stratigraphy, paleontology, and paleoecology of the great Permian reef complex. Illustrations of reef lithology and some of the fossils are included.

Northrop, S. A., 1962, "New Mexico's Fossil Record." *New Mexico Quarterly,* v. 32, no. 1–2, 74 pp.

A fine but unillustrated summary of New Mexico fossils, filled with information not only about paleontology, but also about New Mexico stratigraphy and the early history of paleontological study in the state.

Reeside, J. B., Jr., 1924, *Upper Cretaceous and Tertiary Formations of the Western Part of the San Juan Basin, Colorado and New Mexico.* U.S. Geological Survey Professional Paper 134, 70 pp.

Summary of the Upper Cretaceous and Lower Tertiary stratigraphy, with lists of fossils and localities, of northwestern New Mexico.

Romer, A. S., 1966, *Vertebrate Paleontology.* University of Chicago Press, 468 pp.

The most widely used textbook on vertebrate paleontology.

Shimer, H. W. and Shrock, R. R., 1944, *Index Fossils of North America.* John Wiley & Sons, 837 pp.

Short descriptions and thousands of illustrations of most of the conspicuous North American invertebrate fossils. Some of the names of fossils have changed since the book was published, but it is still an invaluable and detailed source of information about fossils of all ages found in North America.

REFERENCES

Spinar, Z. V. and Burian, Z., 1972, *Life before Man.* American Heritage Press, 228 pp.
  Most of this book is made up of paintings (over 150 of them) of extinct animals and plants from the Precambrian to the present. World-wide in its scope, the volume includes several illustrations of organisms found in New Mexico rocks. This is the best introduction I know to the actual appearance of extinct organisms.

Sutherland, P. K., and Harlow, F. H., 1973, *Pennsylvanian Brachiopods and Biostratigraphy in Southern Sangre de Cristo Mountains, New Mexico.* New Mexico Bureau of Mines & Mineral Resources Memoir 27, 173 pp.
  This very detailed monograph describes and illustrates the most numerous and characteristic Pennsylvanian fossils—brachiopods—from the largest area of Pennsylvanian outcrops in the state. Many of the species mentioned are widely distributed within New Mexico.

Tidwell, W. D., 1975, *Common Fossil Plants of Western North America.* Brigham Young University Press, 197 pp.
  Excellently illustrated comprehensive summary of fossil plants of the West.

Trimble, S., 1980, "From out of the Rocks." Life History Program Advisory Committee and Museum of Northern Arizona, 32 pp.
  A short but concise and well-written introduction to New Mexico fossils, aimed at a high-school-level audience.

Whitaker, G. O., 1965, *Dinosaur Hunt.* Harcourt, Brace & World, Inc., 94 pp.
  The story of the discovery of *Coelophysis*, the little dinosaur from Ghost Ranch, with description of how paleontologists work in the field, and how fossils are discovered and excavated.

INDEX

Italicized page numbers refer to illustrations.

Abiquiu, 78, 125, 196
Abo Formation, 78, 125
Acanthoceras, 159, *160*
acanthodians, 62, 96, 101, *118*
acid etching, 30, 123
Adelophthalmus, 113
Aden Crater, 201
"Age of Fishes," 101
"Age of Mammals," 176
"Age of Reptiles," 171
agnathans, 62, 87, 89, 101
airbrasive unit, 30–31
Akanthosuchus, 178, *179*
Alamogordo, 113
Alamosaurus, 168
Albuquerque, 113, 121, 147, 158, 183, 189, 199, 200
algae, *38*, 60, 73, 74–75, 87, *92*, 96, 121, 150
Allognathosuchus, 178, *178*
Allosaurus, 147
Almagre Arroyo, 183
amber, 7
American Indians. *See* Indians; Paleo-Indians
American Museum of Natural History, 136, 191, 196
ammonoids, 4, 22, 53, *208*; Cretaceous, 150, *154*, *155*, 158–59, 172; Devonian, 98, *100*, 101; Jurassic, 141; Mississippian, 102, *107*; Pennsylvanian, 105, *118*, 120; Permian, 129, *133*; Triassic, 134
amphibians, 63, *64*–65; description of, 66; Devonian, 101; Miocene-Pliocene, 195; Mississippian, 102;
Pennsylvanian, 111, 121; Permian, 125, 126–27; Triassic, 134, 138
angiosperms, 76, 77–78, *80*, 179, 185
animals: first land, 96; general features of, 37–41; kingdom of, 34, 36. *See also* invertebrates *and* vertebrates
ankylosaurs, 68, 169
annelids, 49–50, *49*
Antiquatonia, 114
Antrodemus, *145*, 147
Apache Canyon (Quay County), 137
Apatosaurus, 144, 147
Appalachian Mountains, 95
arachnids, 56
Araucarioxylon, 138, 140
archaeocyathids, 37, 87
Archimedes, 102, 105
archosaurs, 66, 69, *70–71*, 134
Arizona, 77, 78, 138
Armstrong, A. K., 105
arthropods, *38*, 40–41, *57*, 96, 102; arachnids, 56; crustaceans, 56; eurypterids, 56; insects, 58; trilobites, 55
artiodactyles. *See* ungulates
asteroid impact, 173–74, 207
Atlantic Ocean, 101
Australia, 112, 121

Baca Formation, 183, 190
bacteria, 36, 73, 87
Baculites, 159
Baldwin, David, 125, 126, 137, 176
Battleship Rock, 113
bedding planes, 9, 26, 31, 61
behavior, in the fossil record, 3, 7, 10

217

INDEX

benthonic organisms, 37, *38*, 134
Bernalillo, 147
Big Hatchet Mountains, 89, 102, 148, 150
"bird-hipped" dinosaurs, 68, 144, 147
birds: Cretaceous, 163; description of, 67, 68, 69; Eocene, 187, 189; Jurassic, 147; Miocene-Pliocene, 194; Pleistocene, 206
biology, 206
biostratigraphers, 21
Bisti, 164, 168
Blackwater Draw Site, 199, 204–5
Blanco Arroyo, 183
blastoids, 58, 102, 105, *108*
Bliss Sandstone, 89
Bloomfield, 179
Bluewater Lake, 124
bone, 9, 62
"bony fish," *38*, 63, 101, 102, *118*, 127, 141, *156*
borings, 58
brachiopods, 21, *38*, 40, *48*; Cambrian, 87, *88*, 89; Cretaceous, *152*, 161; description of, 47, 49; Devonian, 98, 99, *100*, 101; Mississippian, 102, 105, *106*, *107*; Ordovician, 89, *93*, 95; Pennsylvanian, 105, 114, *115*, *116*; pentamerid, 98; Permian, 123, 127, *130*, *131*; Silurian, 96, *97*, 98; Triassic, 134
"brittle stars," 60
Brontosaurus, 144, 147
bryozoans, *38*, 40, *46*; Cretaceous, 161; description of, 46–47; Devonian, 98, 99, 101; Jurassic, 141; Mississippian, 102, *106*; Ordovician, 89, *92*, 95; Pennsylvanian, 114, *115*; Permian, 121, 129, *130*; Triassic, 134
Bureau of Land Management, 25, 26, 148, 176
Burro Mountains, 162
burrows, 10, *10*, *49*, 50, 56, 127, 161

Caballo Mountains, 89, 95
Calamites, 75, 77, 109, 120
calcite, 8, 14
calcium phosphate, 9, 61, 62
Camarasaurus, 68, 144, *146*, 148
Cambrian Period, 17, 37, 74, 84, 95; brachiopods of, 49; events of, 89; fossil assemblage for, 87, *88*, 89; graptolites of, 61; map for, *90*; trilobites of, 56; vertebrates of, 62
Camelops, 203
camels, *15*, 193, 200
Camptosaurus, 144
Caninia, 120
Carboniferous Period, 17, 75, 76
carbonization, 9
Carlsbad Caverns, 124–25
carnivores, 34, 39, 66, 126, 180, 181, 186, 190, 194, 203
carnosaurs, 147, 148, 169–70
casts, 9–10, 120
catastrophic events, 173
cellulose, 8
Cenozoic Era, 17, 73, 175, 176
cephalon, 55
cephalopods, 53, *54*, 55, 141. *See also* ammonoids *and* nautiloids
ceratopsians, 68, 168–69
Cerro de Cristo Rey, 148
Chaco Canyon, 164
Chaetetes, 114, *120*
"chain coral," 98
champsosaurs, 178
chemical fossils, 3, 7, 10
Chinle Formation, 78, 138
chitin, 41
chitons, 55
chlorophyll, 10, 35
chondrichthyans, 62–63
chordates, 34, 62
Cimarron River, 136
clams, 47, 50, 51
class, 34
classification: development of the system of, 33–34; general features used for, 37–41; invertebrate, 41–62; main divisions in, 35–37; nomenclature for, 35; plant, 73–81; use of body plan for, 37; vertebrate, 62–73
Clovis, 73, 204
"club mosses," 75
coal, 9, 22, 75, 78, 109, 111, 120, 164, 176, 182
Cobre Canyon, 137
coelenterates, *44*, 45–46. *See also* corals

218

INDEX

Coelophysis, 66, 68, 136, 137, *140, 141*
coelurosaurs, 147, 169
Coilopoceras, 158, *160, 208*
colonial species, 40, *40*, 45, 46
Colorado, 69, 113, 176, 182
coloration in fossils, 8, 47, 77
Composita, 114
Conchidium, 98
concretions, 159
condylarths, 180–81, 185, *188*
conglomerate, 11, 14
conodonts, 22, 111, 129
conularids, 46
Cooks Range, 89
Cope, Edward Drinker, 125, 137, 176, 183, 185, 191
copper, 125
coprolites, 10, 203
corals, *38*, 40, *44*, 53, 74; colonial, 95, 96, 98; Cretaceous, 46, 150, *156*, 161; described, 45; Devonian, 98, 99, 101; Jurassic, 141; Mississippian, 102, 105, *106*; Ordovician, 89, *91*, *92*, 95, 96; Pennsylvanian, 105, 114, *115*, 120; Permian, 121, 129; scleractinian, 141, 150; Silurian, 96, 97, 98; Triassic, 134. *See also* reefs
Cordaites, 77, 77, 109, *113*
Coryphodon, 185, *187, 188*
Coyote, 125
creodonts, 180, 181, 185, 190
Cretaceous Period, 17, 21, 121, 141, 143, 176, 178, 179, 182, 185; brachiopods of, 49; burrows of, 50; corals of, 45, 46; crustaceans of, 56; echinoderms of, 60; events in, 148, 172–75; forams of, 41, 42; forests of, 164–66; fossil assemblage for, marine, 148, 150, *151, 152, 153, 154, 155, 156, 157*, 158, 161–62, vertebrate, 163, 168–70, 171; map for, 149; molluscs of, 50, 53, 55; panorama for, 158; plants of, 77, 78, 81; rocks of, 21, 84, 158–59, 161; sharks of, 63
crinoids, 21, *38*, 58, 89; Jurassic, 141; Mississippian, 102, 105, *107*; Ordovician, *91*; Pennsylvanian, 105, 114; Permian, 129; Triassic, 134
crocodilians, 66, 147, 165, 178, 185
crustaceans, 55, 56, *57*, 121, 161

Cuba, 78, 163, 164, 179
cyclothems, 112
Cyrtospirifer, 101
cystoids, 60

Dalmanella, 98
Datil (area), 183, 190
dating: absolute and relative age, 17; by specific episode, 21; geological time for, 16–20; radiocarbon, 205; radiometric, 17
Dentalium, 150
deposition, 5, 13. *See also* sedimentary rocks
Derbyia, 114
Devonian Period, 17, 84, 102, 105; ammonoids of, 53; events in, 98, 101; fossil assemblage for, 98, *99, 100*; map for, 103; plants of, 75; rocks of, 101; vertebrates of, 62, 63, 66
Diadectes, 127
diatom, *38*
Diatryma, 187, 189, *190*
Dimetrodon, 126, *126*
Dinosaur National Monument, 144
dinosaurs, 66, 68, 69, 136, 137, 144, 147, 148, 166, 168–70, *169, 170, 171, 172*, 174–75, 182, 189
Diplodocus, 144
Dockum Group, 138
"duck-billed" dinosaurs, 68, 168
Dulce, 162

Eagle Nest, 113
earth, age of the, 16. *See also* dating
earthworms, 39, 49
echinoderms, 58, *59*, 60; Cambrian, 87; Devonian, 98, 101; Mississippian, 102, 105, *108*; Ordovician, 89, 95; Permian, 121, 129; Silurian, 96; symmetry of, 39–40. *See also* blastoids *and* crinoids
echinoids, *38*, 60, 109, *115*, 141, 150, *156*
eggs, 10, 66, 76, 111
El Cobre Canyon, 125, 138
Elephant Butte Reservoir, 170
elephants, 187, 193, 199
El Paso, 87, 150
El Paso Group, 95

219

INDEX

Endothyra, 42, 105
environments: Cambrian, 87; collecting area, 29; Cretaceous, 148, 163–65; depositional, 84, 86; Devonian, 102; for fossilization, 12–14, 15, 16; Jurassic, 140, 141, 143; Miocene-Pliocene, 195–96; Mississippian, 102; of geologic time in table form, 18–19; Oligocene, 190; Ordovician, 89; Paleocene, 182–83; Pennsylvanian, 109; Permian, 121, 123, *123*, 124; Pleistocene, 199; reconstructing, 21–22; Silurian, 96; terrestrial, 125
Eocene Epoch, 84, 182; fossil assemblage for, 183, 185; mammals of, 73, 186–87, 189, 190; map for, 184; plants of, 78; rocks of, 183, 189; vertebrates of, 69, 73
Eohippus, 185–86
Ephedra, 77
erosion, 13. *See also* sedimentary rocks
Eryops, 126, *128*
Española, 73, 196
eukaryotes, 36
eurypterids, 56, *57*, 105, 111, 121, 129
evolution, 21, 35; convergent, 37, 53; plant, 78
Exogyra, 159
extinctions, 17, 20, 21, 207; ammonoid, 53, 159; amphibian, 134; arthropod, 55, 56, 58; concept of, 4; Cretaceous, 150, 159, 172–75; Devonian, 102; dinosaur, 68, 174–75, 182, 189; Eocene, 185; graptolite, 61, 98, 102; mammal, 73; of a phylum, 37, 87; Paleocene, 182; Permian, 127, 129; Pleistocene, 203, 205–6; Triassic, 134; vertebrate, 62, 69

families, 34
Farmington, 69, 78, 164
feces, 10
feeding behavior, in the fossil record, 37, 39
ferns, 9, 21, 76, 78, 109, 138, 158
fish, 20, 21, *64–65*; Cretaceous, 159, 161; Devonian, 98, 101; Mississippian, 102; Ordovician, 89; Paleocene, 178; Pennsylvanian, 111, 121; Silurian, 96; Triassic, 134, 136

floating organisms, 37, *38*
Florida, 14, 165
Flower, Rousseau, 95, 96
flowering plants, 164. *See also* angiosperms
Folsom Site, 73, 199, 204, 205
footprints, 10, 15, *15*, *129*, 136
foraminifers, (forams), 4, *38*, 41; Cretaceous, 150, 161, 172; Mississippian, 102, 105, *107*; Ordovician, 89; Pennsylvanian, 109, 114. *See also* fusulinids
forests, 75, 78, 98, 102, 109, 111, 125, *164*
Fort Wingate, 78, 137, 138
fossil assemblages, 16
fossilization: affected by sediment, 14; body-type, 7–10; carbonization, 9; chemical, 7, 10; concept of, 4; environments for, 12–14; petrifaction, 8–9; process and chances for, 5–6; recrystallization, 7–8; replacement, 8, 125; steinkerns, casts, and molds, 9; trace, 7, 10; unaltered-type, 7
fossils: appeal of, 22–23; as documentation, 20–21; assemblages of, 16, 17; body, 7–10; chemical, 7, 10; cleaning, 30–31; collecting access to, 25–26; collecting ethics for, 28–30; collecting equipment for, 26–28; collecting preparation for, 23–26; displaying, 31; distribution of, 84–86; economic use for, 22; general features of, 37, 39–41; identifying, 27, 28, 29, 30, 37, 83; literature on, 24; ownership of, 25, 26; specimen preparation of, 26, 30–32; trace, 7, 10; where to find, 11–16
Fra Cristobal Range, 196
Franklin Mountains, 87, 89
Frick, Childs, 191
Fruitland Formation, 164–72 passim, 176
fusulinids, 42; Pennsylvanian, *109*, 114, *115*; Permian, 121, 125, 129, *130*
Fusselman Dolomite, 96, 98

Galisteo, 125, 158
Galisteo Formation, 183, 189–90
Gallina, 137, 183, 191
garfish, 165, *166*, 178

INDEX

gas (natural), 124
gastropods: Cretaceous, 150, *153*, 159, 161; description of, 50, *51*; Devonian, 98, *100*, 101; Jurassic, 141; Mississippian, 102, *107*; Ordovician, 89, *93*, 95; Pennsylvanian, 35, 105, 114, *117*; Permian, 123, *132*; Silurian, *97*; Triassic, 134
genetic relationships, 20, 35
genus, 34, 35
geochemists, 10
geode, 9
geological time, 16–20; development of scale for, 16; divisions for, 17, *18–19*; major events of, 17; North American variations on, 17
geology, 5, 21
Gesner, Konrad von, 4
Ghost Ranch, 66, 136, 137
gomphotheres, 193, 196
Gondwanaland, 112
Granocardium, 159
Grants, 124, 147, 148
graptolites, *60*, 61; Devonian, 98; Mississippian, 102; Ordovician, 89, *92*, 95; Silurian, 96
Great Barrier Reef, 121
Great Salt Lake, 144
Gregg, Josiah, 190
ground sloth, 7, *204*
ground water, 8, 9
Guadalupe Mountains, 121, 124, 199
gymnosperms, 76, 77, *80*, 109, 138
gypsum, 143, 144

hadrosaurs, 168
Halysites, 98
Helicoprion, 125
Hemichordata, 61
Hemicyon, 193
Hermosa, 190
Hillsboro, 101, 190
holothurians, 60
Homo sapiens, 199
Hormotoma, 98
"horn corals," 45
horned dinosaurs, 168–69
horse, 73, 185–86
Hyracotherium, 185–86, *189*

ichthyosaurs, 68, 143
Idonearca, 159

igneous rock, 15
Indians, 4, 25. *See also* Paleo-Indians
Inoceramus, 159
Insectivora, 180, 185
insects, 7, 41, *57*, 58, 111, 121, 179
invertebrates, 7, 36, 83; annelid, 49; arthropod, 55; brachiopod, 47; bryo-.zoan, 46; Cambrian, 87; coelenterate, 45; conodont, 61; Cretaceous, 172; Devonian, 98; echinoderm, 58; extinct, 102; flying, 58; graptolite, 61; Jurassic, 143–44; mollusc, 50; Ordovician, 89, 95; Pennsylvanian, 121; Permian, 121, 123, 125, 129; porifera, 42; protozoan, 41; silicified, 30; Silurian, 96; skeletons of, 40
Isleta Cave, 199

Jefferson, Thomas, 4
jellyfish, *38*, 45
Jemez area, 105, 120, 199
Jemez Pueblo, 113
Jemez Springs, 8, 125
Jornada del Muerto, 162
Jurassic Period, 22, 182; dinosaurs of, 168; events of, 143; map for, *142*; marine fossil assemblage for, 141, 143; plants of, 78; rocks of, 22, 84, 140, 143, 144, 147; terresterial fossil assemblage for, 144, 147, 148; vertebrates of, 63, 66, 68, 69
Juresania, 114

Kansas, 159, 163
kingdoms, 34, 35–36
Kirtland Formation, 164–72 passim
Kritosaurus, *169*, *174*, 175
Kruschevia, 96

Lake Valley, 58, 101, 102, 105
Laminatia, 101
Lamy, 46, 66, 134, 137, 161
Las Cruces, 7, 199, 201
Las Vegas, 105
Lepidodendron, 75, 77, 109, *112*, 120
life processes, 20
limestone, 11, *13*, 14, 30, 42; Cretaceous, 150; Jurassic, 143; Mississippian, 102; Ordovician, 95; Pennsylvanian, 109, 114; Permian, 121, 123, 124; Silurian, 98
Limnoscelis, 126, *127*

221

INDEX

Linnaeus, 34
Linné, Carl von, 33
lithification, 6
Little Hatchet Mountains, 148, 199
Lopha, 150, 159
Los Lunas, 56, 58, 120, 196
Lucas, Spencer, 185
Lucero Mesa, 78, 120

McRae Formation, 170
Magdalena Mountains, 102
mammals, 20, 66, 67, 68, 72, 78; archosaurs, 70–71; bones of, *180, 181, 194, 195*; definition of, 34; description of, 69, 72, 73; Cretaceous, 171; Cenozoic, 176, 178, 180, 181, 182, 193, 199; Triassic, 136
mammoths, 5, 7, 199–200, *200, 202*
Mancos Shale, 163
Manticoceras, 98
Manzanita Mountains, 121
Manzano Mountains, 62, 78, 113
maps, 23–24, *24, 25*, 25, 28
Marcon, Jules, 148
marine organisms, 13. *See also* invertebrates
Marsh, Othniel C., 125
mastodons, 199–200, *201*
matrix around fossils, 30–31
Mesa Lucero, 78, 120
Mesa Redonda, 137
Mesozoic Era, 17, 98, 171; burrows of, 50; marine fauna of, 143; molluscs of, 53; plants of, 77; vertebrates of, 61, 66, 68, 73
metamorphic rock, 16, 87
metazoans, 36, 87
Metoposaurus, 134, *136, 137*
Millerella, 109
millipedes, *57*, 58
Mimbres Mountains, 89
mineral precipitation, 7–9, 11, 124
Miocene-Pliocene epochs: fossil assemblage for, 193–96; map for, *192*; molluscs of, 51, 53; plants of, 81; rocks of, 191, 193, 196; vertebrates of, 69; 73
Mississippian Period, 17, 109; fossil assemblage for, 102, *106, 107, 108*; echinoderms of, 58, 60; graptolites

of, 61; map for, 103; molluscs of, 50; panorama for, *104*; plants of, 75; 78; protozoans of, 42; rocks of, 102, 105
modes of life, 37, *38, 40, 64–65,* 67
molds, 9, *10*
molluscs, 37; Cambrian, 87; Cretaceous, 150, *151, 152, 153*, 166; description of, 50, *51, 52, 53, 54,* 55; Devonian, *100*; Miocene-Pliocene, 196; Paleocene, 178, 178–79; Pennsylvanian, 114; Permian, 121, *132*; Pleistocene, 206; Silurian, *97*; Triassic, 134. *See also* cephalopods, gastropods, pelecypods
Monera, 36, 73
Montoya Group, 95
Morrison Formation, 68, 144, 147, 148, 168
mosasaurs, 68, *161*, 162
"moss animals," 46
mother-of-pearl, 22, 166
mountain building, 95, 101
Mud Springs, 89
multicellular animals, 17, 36, 42
multituberculates, 182
mummification, 7

Nacimiento Formation, 176, 178, 179–80
Nacimiento Mountains, 78, 113, 114
Nambe (area), 195
Native Americans. *See* Indians, Paleo-Indians
nautiloids, *38*, 53, 159; Cretaceous, *154*; Devonian, 98, *100*, 101; Mississippian, 102, *108*; Ordovician, 89, *91, 94*, 95; Pennsylvanian, 105, *118*, 120; Permian, *133*; Silurian, 96
Neithea, 150
nektonic organisms, 37, *38*
Neospirifer, 114
Neuropteris, 109
Newberry, J. S., 138
New Mexico Bureau of Mines and Mineral Resources, 23, 95, 211
New York, 191

Ogallala Formation, 196
oil in reefs, 150
Ojo Alamo Spring, 168

INDEX

Oligocene Epoch, 182; 193; fossil assemblage for, 190
O'Neill, Mike, 178
On Fossil Objects, 4
Ophiacodon, 126, *128*
Ophiomorpha, *160*, 161
ophiuroids, 60
Orbitolina, 150
orders, 34
oreodonts, 193
Ordovician Period, 17, 62, 89, 96, 98; arthropods of, 56; brachiopods of, 49; corals of, 45; echinoderms of, 58; events in, 95; fossil assemblage for, 89, *91*, *92*, *93*, *94*, *95*; graptolites of, 61; map for, 90; molluscs of, 50, 53, 55; panorama of, 91; rocks of, 95; trilobites of, 56
Organ Mountains, 89, 199
Oscura Mountains, 125
osteichthyans, 63
ostracods, 56, 97, 111, 143, 161
Ostrea, 150, 159
oysters, 51, 53, 148, 150, 159

Paleocene Epoch, 84, 171, 185; fossil assemblage for, 176, 178–83; map for, 177; petrified wood of, 58; plants of, 78, 81, 179, 182; vertebrates of, 69, 73
Paleo-Indians, 73, 199, 204, 205–6
paleontologists, 24–25, 27, 28, 29, 30, 83, 148, 211
paleontology, 22; subdivisions of, 5
Paleozoic Era, 17, 22, 61, 89, 95, 98, 102, 125, 127, 134; arthropods of, 56; brachiopods of, 47, 150; bryozoa of, 47; burrows of, 50; coelenterates of, 45, 46; echinoderms of, 58, 60, 105; events of, 111, 129; fossil record for, 89; molluscs of, 53, 150; plants of, 77; vertebrates of, 66; world configuration for, 101
Pangaea, 111, 129
Parasaurolophus, *167*, 168
parasitism, 39
Pecos River, 105, 113
Pecos Valley, 125
pectinids, 53
pelecypods, 38, 47, 52; Cretaceous, 150, *151*, *152*, 159, 172; description

of, 51, 53; Devonian, 98, *100*, 101; Jurassic, 141; Mississippian, 102; Ordovician, 89, 95; Pennsylvanian, 105, 111, 114, *117*; Permian, 123, *132*; Silurian, 97; Triassic, 134, 137, *139*
Pennsylvanian Period, 17, 21; arthropods of, 56, 58; events of, 111–12; forams of, 125; fossil assemblage for, 105, 109, 111, 114, *115*, *116*, *117*, *118*, *119*, 120–21; fossil coloration for, 8; insects of, 58; map for, 110; molluscs of, 35, 55; panorama for, 114; plants of, *9*, 77, 78, 81, 125; rocks of, 21, 84, 112–13; vertebrates of, 62, 63, 66
Pentaceratops, *168*, 169
Pentamerus, 98
perissodactyls, 180
Permian Period, 17, 62, 112, 134, 150, 175; arthropods of, 56; boundary with Pennsylvanian, 121; brachiopods of, 49; corals of, 45; events of, 121, 124, 129; fossil assemblage for, *130*, *131*, *132*, *133*; marine fossil assemblage for, 121, 123–25; map for, 122; molluscs of, 50, 55; panorama for, 124; plants of, 75, 78; protozoans of, 42; reef of, 30; 43; *123*; rocks of, 121, 123, 124; sponges of, 43; terrestrial fossil assemblage for, 125–127, 129; vertebrates of, 63, 66, 68
petrifaction, 8–9, 74
Petrified Forest National Park, 77, 78, 140
petrified wood, 8, 22, 58, 78, *163*; Cretaceous, 159, 164; Eocene, 185, 190; Miocene-Pliocene, 195; opalized, 196; Paleocene, 179
petroleum, 10, 22, 124. See also oil in reefs
Pholadomya, 159
Pholidophorus, *143*
photography, 29
photosynthesis, 35
phyla, 34, 36–37
phytoplankton, 173
phytosaurs, 68, 134, 138
Pinna, 159
Placenticeras, 158
placoderms, 62, 101

223

INDEX

planktonic organisms, 37, *38*, 41, 42, 46, 61, 161
Planoproductus, 101
plants, 9, 84, 173; algae, 74, 75; angiosperm, 77, *80*, 164; Cretaceous, *157*; description of, 35, 73–81; Devonian, 98; groups of, 74; gymnosperm, 77, *80*; land, 75, 96; lower vascular, 75–76, 78, 79, 102, 109; Jurassic, 147; Miocene-Pliocene, 196; Mississippian, 102; New Mexico fossil record for, 78–81; Paleocene, 178, 179, 182; Pennsylvanian, 77, 109, *119*, 120; Permian, 125; Silurian, 76; Triassic, 134, 137, 138, *139*; vascular, 75
plate tectonics, 22. *See also* continents
Platystrophia, 95, 98
Pleistocene Epoch, 7, 196; fossil assemblage for, 199, 200–201, 206; glaciation in, 112, 196, 206; molluscs of, 51, 53; plants of, 81; vertebrates of, 69, 73, 200–201
plesiosaurs, 68, 143, 162
Pliocene Epoch, *197*. *See also* Miocene-Pliocene epochs
Pojoaque Pueblo, 191
Poleo Creek, 125
polyplacophorans, 55
porifera, 42, *43*. *See also* sponges
Portales, 204
potash, 124
Precambrian Period, 17, 36, 86; fossil assemblage for, 87, plants of, 73, 74; rocks of, 87
primates, 181
productoids, 47
prokaryotes, 36
Protista, 36
protozoans, 36, 41, *42*. *See also* foraminifers
pterosaurs, 66, 68, 147, 172
Ptilodus, 183
Punctospirifer, 114
Pycnodonte, 159
pygidium, 55
Pyramid Mountain, 148

quartz, 8
Quaternary Period, 176; events of, 196, 205–6; fossil assemblage for, 199, 200, 201, 203, 206; map for, 198; Paleo-Indians of, 204–5; rocks of, 199
Quay County, 137

Raton, 78, 113, 158, 162, 164, 182, 183
rays, 62, 143, 161–62; teeth of, *156*
Recent Epoch, 196; the present, 206–7
Receptaculites, 95–96
recrystallization, 7, 16
reefs, 14, 22, 45, 74, 96; Cretaceous, 150; Devonian, 98; Mississippian, 102; Permian, 121, *123*; Silurian, 98
reptiles; Cretaceous, 162, 166; description of, 66, 67, 68–69; Eocene, 185; Jurassic, 143, 147; Miocene-Pliocene, 193, 194; Paleocene, 178; Pennsylvanian, 111; Permian, 125, 126, 134; Triassic, 134
Rhipidomella, 105
Rigby, J. Keith, Jr., *146*, 147
Rincon Hills, 81
Rincon Mountains, 183
Rio Grande, 196
Rio Puerco, 60, 158
rodents, 182, 186, 190
Roswell, 113
rudistids, 53

Sabalites, 162
"sabre-toothed" cats, 194, 203
Sacramento Mountains, 89, 101, 102, 113, 123
"sail-backed" reptile, 126
salt, 124, 144
San Andres Mountains, 60, 89, 95, 101, 102, 123
San Augustin Plains, 199
Sandia Cave, 199
Sandia Mountains, 21, 105, 113, 158, 199
San Diego Canyon, 113
sandstone, 11, *12*, 30, 50, 60, 63, 78; Cambrian, 89; Cretaceous, 161; Eocene, 183, 185; Pennsylvanian, 120; Permian, 124, 125
Sangre de Cristo Mountains, 78, 105, 113, 127
San Ildefonso Pueblo, 191
San Jon, 137
San José Formation, 183, 185, 186, 189

INDEX

San Juan Basin, 81, 158, 163, 165, 170, 176, 182, 183, 185, 186
San Marcial, 196
San Mateo Mountains, 190
Santa Cruz (area), 195
Santa Fe, 16, 73, 113, 183, 189, 191, 193, 196
Santa Fe Group, 193–94, 195–96
Santa Rosa, 78, 137, 138
San Ysidro, 147
sauropods, 68, 144, 147, 168
Scaphites, 158
scaphopods, 55, *118*, 120, *132*, 150, *154*
Schizophoria, 105
scientific names, 35
scleractinians, 45, *156*
scorpions, 56
scutes, 134, 165, *166*
"sea cucumbers," 60
"sea lilies," 58
seasonality, 172
sea urchins, 60
sedimentary rocks, 3, 6, 21, 29, 74; deposition for, 112, 144, 199; formation and types of, 11; fossil occurrence in, 12–14; interval names for, 16–17; New Mexican sequence of, 84, *85*; variation in thickness of, 84
seed ferns, 109
seeds, 76
segmented worms, 49
septa, 45, 53
sessile organisms, 13
shales, 11, *11*, 14, 30, 31, 50, 61, 78; Eocene, 183; Devonian, 101, 102
sharks, 37, 62, 64, 65, 101; Cretaceous, 63, *156*, 159, 161; Jurassic, 141; Mississippian, 102, 105; Pennsylvanian, *118*; Permian, 125, 127, *133*
shells, 7, 47; buoyancy of, 53; calcareous, 41, 49, 50, 51, 53, 56; phosphatic, 47; remade, 8
shrimp, 56, 111
silica, 8, 78
Silurian Period, 17, 22, 84, 101, 102; corals of, 45; fossil assemblage for, 96, 97, 98; map for, 90; panorama for, 91; plants of, 74, 75, 76; rocks of, 96, 98
Silver City, 89, 101, 162
single-cell organisms, 20, 36, 41, 125

skeletons, 6, 7, 37, 62; chitinous, 41, 55; colonial, 40, *40*, 45, 46–47, 61; comparisons of types of, 40–41; conodont elements of, 61; echinoderm, 58; endurance of, 13–14; phosphatic, 46; plant, 74; protozoan, 36; replacement of, 8; sponge, 43; use for classification of, 36
Sly Gap, 101
snails, 50, 159
Socorro, 78, 125, 158, 183, 196
species, 21, 33, 34–35
Sphenacodon, 126, *128*
spicules, 43, 60
spiders, 56
"spiny-skinned" animals, 58
sponges, 42, *43*, 45, 87, *92*, 95, 121, *130*
spores and pollen, 5, 81
squids, 50
Stegosaurus, 68, 144, *145*, 147
steinkerns, 9, 10, *10*, 114
Streptelasma, 98
stromatolites, 86, 87, 96
stromatoporoids, 45, *92*, 96
strophomenids, 101
structure, 3, 7; classification by, 34; feeding behavior influence on, 37, 39, 78; mode of life influence on, 37; plant, 75, 77; symmetry in, 39–40. *See also* skeletons
substrate-bound organisms, 37, *38*, 49, 58
sunlight, 173–74
supernova, 129
swimming organisms, 37, *38*, 53, 56
symmetry, 39–40, *39*, 47, 58, 60

taeniodonts, 185
Taos, 35, 113
tectonic events, 84, 134. *See also* continents
Teleoceras, 193
Teleodus, 191
teleosts, *64–65*, 159
Tellina, 159
Tempskya, 158
terrestrial organisms, 13
"terror crane," 187, 189
Tertiary Period, 84, 176, 187, 193; molluscs of, 50, 53; plants of, 78
Texas, 30, 121, 123, 125, 196

225

INDEX

Texigryphaea, 148, 150
Tierra Amarilla, 183
tillodonts, 185
titanotheres, 189, 190
Todilto Formation, 143
Todilto Lagoon, 144
Tonuco Formation, 89
tortoises, 194–95, *195*
trace fossils, 7, *10*, *49*, 50, 120, 127, 161
transgressions, 172; Cambrian to Ordovician, 89; Cretaceous, 148; Devonian, 101; Jurassic, 143
trees, 182, 190
Triassic Period, 129; dinosaurs of, 144; events of, 134, 136; fossil assemblage for, 134, 136, 137, *139*; map for, 135; panorama for, 138; plants of, 78; rocks of, 134, 138; vertebrates of, 66, 68, 69
trilobites, 55–56, *57*; Cambrian, 87, *88*, 89; Devonian, 98, *100*, 101; Mississippian, 102, *107*; Ordovician, 89, *91*, *95*; Pennsylvanian, 105, *118*; Permian, 129; Silurian, 96
Trinidad, 182
Tucumcari, 78, 137, 138, 148
Tularosa, 113
Turritella, 161
turtles, 68, 147, *165*, 165, *186*
tusk shells, 55
Typothorax, 134
Tyrannosaurus, 66, 169, *173*

unaltered fossils, 7
ungulates, 180–81, 185, 187, 189, 190, 193
uniformitarian dictum, 207
Union County, 136
United States Geological Survey, 23, 24, 28, 105
University of New Mexico, 147, 176, 195
uranium, 22, 147, 148
Utah, 69, 144

Valles Caldera, 199
valves, 47, 53
Vermejo Park, 78, 164
vertebrates, 14, 26, 27, 29, 36, 62–73, *70–71*, 83–84; acanthodian, 62; agnathan, 62; amphibian, 66; aquatic, *64–65*; bird, 67, 69; Cambrian, 87; chondrichthyan, 62; Cretaceous, 161, 163, 171; Devonian, 98, 101; Eocene, 183; flying, 66, 68, 69; mammal, 67, 69, *72*; Miocene-Pliocene, 193; Mississippian, 102; Oligocene, 190; Ordovician, 89; osteichthyan, 63; Permian, 125; placoderm, 62; Pleistocene, 199; reptile, 66, 67; Silurian, 96; skeletons of, 40; Triassic, 134, 136, 140
Virgiana, 98
volcanic activity, 6, 199
volcanic rock, 15, 87
Volutomorpha, 161

Walchia, 109, *111*
water-putty, 31
"water scorpions," 56
Wheeler Geological Survey, 183

Yale University, 125

Zamites, 138
Zuni Mountains, 113, 124